Covert Modality in Non-finite Contexts

Interface Explorations 8

Editors

Artemis Alexiadou
T. Alan Hall

Mouton de Gruyter
Berlin · New York

Covert Modality in Non-finite Contexts

by

Rajesh Bhatt

Mouton de Gruyter
Berlin · New York

Mouton de Gruyter (formerly Mouton, The Hague)
is a Division of Walter de Gruyter GmbH & Co. KG, Berlin.

∞ Printed on acid-free paper which falls within the guidelines
of the ANSI to ensure permanence and durability.

Library of Congress Cataloging-in-Publication Data

Bhatt, Rajesh, 1973–
Covert modality in non-finite contexts / by Rajesh Bhatt.
p. cm. – (Interface explorations ; 8)
Includes bibliographical references and index.
ISBN-13: 978-3-11-017952-1 (cloth : alk. paper)
ISBN-10: 3-11-017952-0 (cloth : alk. paper)
1. Modality (Linguistics) 2. Grammar, Comparative and general – Relative clauses. I. Title.
P299.M6B48 2006
415'.6–dc22
2006032087

ISBN-13: 978-3-11-017952-1
ISBN-10: 3-11-017952-0
ISSN 1861-4167

Bibliographic information published by the Deutsche Nationalbibliothek

The Deutsche Nationalbibliothek lists this publication in the Deutsche Nationalbibliografie; detailed bibliographic data are available in the Internet at http://dnb.d-nb.de.

Cover design: Christopher Schneider, Berlin.
Printed in Germany.

Contents

Preface

This book is based on my 1999 University of Pennsylvania Ph. D. dissertation. The modal properties of infinitivals continue to be a poorly understood domain of the syntax-semantics interface and I hope that this book will provide a point of departure to future researchers. They are certain to find many open questions within. Since I submitted my dissertation seven years ago, two substantial related works have appeared that I would direct readers of this volume to: Marga Reis' 2003 paper in the Journal of Comparative Germanic Linguistics on the syntactic and semantic properties of German *wh*-infinitives and Valentine Hacquard's 2006 MIT dissertation on the interaction of aspect and modality.

Dissertations, even more than other kinds of academic endeavors, are collaborative ventures. This one would not have existed but for the advice and support of my advisors and my committee members: Sabine Iatridou, Anthony Kroch, Irene Heim, and Ellen Prince. I would also like to thank my friends and colleagues at the University of Pennsylvania, the Massachusetts Institute of Technology, the University of Texas at Austin, and the University of Massachusetts at Amherst for their support and for their intellectual vigor.

The dissertation that forms the core of this book was dedicated to my family and it makes sense that this book should be too. I dedicate this book to my parents, Shri Kewala Nand Bhatt and Shrimati Savitri Bhatt, my brother, Captain Sanjay Bhatt, and my sister, Dr. Seema Gupta.

Chapter 1
Introduction

1. Issues

Modality embodies an important design feature of natural language - the property of displacement. With modal words, we can talk about what can be but is not, what should be, what is allowed, what is impossible, and what is inevitable. We are no longer limited to reality. Modality gives us access to hope, to despair, and to dreams. In literature and in myth, we see the concept of destiny and attempts of protagonists to change their destiny. Modal reasoning underlies such narratives. Given the conceptual importance of modality, it is not surprising that human languages provide means to express modal thoughts. There are formal languages that lack the means to express modal meanings, but we are unaware of any natural language of this sort.[1]

Modality is typically marked across languages by a class of specialized lexical items called modals. Modal elements do not form a syntactically unified class. They can be verbs, adjectives, adverbs, and nouns across languages. In fact in English itself, we find modal verbs realized as verbs (*can*), adjectives (*possible*), adverbs (*possibly*) and nouns (*possibility*). But in addition, we also find modality in the absence of any specialized lexical modal. We call such modality Covert Modality. Here we interpret modality that is not reliably associated with any overt lexical item. The goal of this book is to investigate the distribution and interpretation of Covert Modality. Covert Modality seems to be dependent upon its immediate syntactic environment in way that overt modality is not. Therefore, studying its properties gives us clues about how the syntax can construct modal meanings and the question of just what it is that makes something have a modal meaning.

This investigation can be seen as furthering two fruitful lines of inquiry in generative grammar. The first line of inquiry involves studying the properties of covert elements. The presence of covert elements is indicated to us by their effects on their environment. Investigation of their properties leads us to a better understanding of the environments that permit them and of universal grammar in general. The second line of inquiry involves making the assumption that the relationship between a structure and its interpretation is always systematic, even when it may not seem so. From this perspective, environments where we seem to interpret something that is not obviously part of the syntactic structure, and vice versa, are particularly telling. This dissertation investigates a systematic subclass of such environments centered around covert modality. Examples of covert modality include *wh*-infinitival complements, infinitival relative clauses, purpose clauses, the *have to* construction, and the *is to* construction:

(1) a. Tim knows [how to solve the problem].
(≈ Tim knows how one/he could/should solve the problem)

b. Jane found [a book to draw cartoons in] for Sara.
(≈ Jane found a book for Sara one could/should draw cartoons in)

c. [The man to fix the sink] is here. (≈ The man whose purpose it is to fix the sink is here)

d. Sue went to Torino [to buy a violin].
(≈ Sue went to Torino so that she could buy a violin.)

e. Bill has to reach Philadelphia before noon.
(≈ Bill must reach Philadelphia before noon.)

f. Will is to leave tomorrow.
(≈ Will is scheduled/supposed to leave tomorrow.)

The interpretation of (1a-f) involves modality; however, there is no lexical item that seems to be the source of the modality. What (1a-f) have in common is that they involve infinitivals.

Of the environments in (1a-f), we will investigate infinitival relatives and infinitival questions. The *have to* construction is discussed in Bhatt (1997). We examine the following aspects of covert modality among others: its source, semantic properties, and distribution. With respect to distribution, we will account for an asymmetry in the distribution of covert modality: all infinitival questions are modal but not all infinitival relatives are.

We will also discuss a case that is the inverse of the environments in (1): an environment where there is a modal word in the structure but not in the interpretation as in (2).

(2) (from Thalberg 1969)

a. In those days, Brown was able to hit bulls-eye three times in a row.
(modal)

b. Brown was able to hit three bulls-eyes in a row. (context: Yesterday, Brown hit three bulls-eyes in a row. Before he hit three bulls-eyes, he fired 600 rounds, without coming close to the bulls-eye; and his subsequent tries were equally wild.)
(non-modal)

This environment is interesting for the same reason as the environments in (1): it involves a non-trivial mapping between the structure and interpretation. How is it possible that (2a) is modal when (2b) is non-modal? Our analysis of this environment will analyze the difference between (2a) and (2b) in terms of a covert modal.

2. Outline and summary

2.1. Chapter 2: The syntax of infinitival relatives

The goal of this chapter is to determine what kinds of structures best represent different kinds of infinitival relative clauses. We begin with a discussion of the syntactic properties of different kinds of infinitival relative clauses. I argue that subject infinitival relatives (e.g. *the man to fix the sink*) differ structurally from non-subject infinitival relatives (e.g. *the book to read*). Subject infinitival relatives are assimilated with reduced relative clauses, while non-subject infinitival relatives are assimilated with non-reduced relative clauses. A structure is proposed for reduced relative clauses that captures the properties of reduced relative clauses in general and subject infinitival relatives in particular. This structure differs from the structure proposed for full relative clauses in that it does not involve a CP projection and A'-movement to the specifier of a CP for predicate-formation. It has the advantage that we do not have to postulate A'-movement of an operator from a non case-marked position.

Non-subject infinitival relative clauses are analyzed as full relative clauses. They involve a CP projection and movement of a relative operator to the specifier of this CP. Infinitival questions and non-subject infinitival relative clauses thus have essentially the same syntax, which involves A'-movement of a phrase to the specifier of a C^0[+rel/wh,+inf]. The infinitival complementizer C^0[+rel/wh,+inf] is identified as the locus of the modality that appears in non-subject infinitival relatives and infinitival questions. Similarities in the behavior of the modality in non-subject infinitival relatives and infinitival questions are captured by assigning similar semantics to C^0[+rel,+inf] and C^0[+wh,+inf]. The details of the semantics of the infinitival complementizer C^0[+rel/wh,+inf] are discussed in Chapter 4.

Some but not all subject infinitival relatives are interpreted as modal (cf. the modal *the man to fix the sink* vs. the non-modal *the first man to walk on the moon*). The proposal that subject infinitival relatives do not involve the modally interpreted infinitival [+rel] C^0 is compatible with the fact that not all subject infinitival relatives are modal. I relate the modality in the modal subject infinitival relatives to a modality that can independently appear in the infinitival clause. Thus the modality in the subject infinitival relative *the book to be read for class* has the same source as the modality in *The book is to be read for class*. This source is internal to the infinitival clause; it is independent of the appearance of the infinitival clause as part of a reduced relative. The infinitival clause in a subject infinitival relative can also be interpreted non-modally in certain environments. The precise characterization of such environments and the properties of such infinitival clauses are discussed in Chapter 3.

The initial analysis proposed for reduced relative clauses is unable to account for data that seem to involve reconstruction of the head into the reduced relative. I show that there are reasons to assume a (head-)raising analysis of relative clauses and suggest a modification to the analysis of reduced relatives in order to permit a head-raising account. Both reduced relatives (e.g. subject infinitival relatives) and full relatives (e.g. non-subject infinitival relatives) involve 'head-raising' of the head NP out of an XP. However, only full relatives involve XP(=CP)-internal A'-movement.

2.2. Chapter 3: Non-modal infinitival relative clauses

This chapter investigates the conditions under which an infinitival relative can receive a non-modal interpretation and the interpretive properties of infinitival relatives that receive such an interpretation.

The syntactic proposal in Chapter 2 is able to explain why only subject infinitival relatives can be non-modal. However, the non-modal interpretation is only available in the presence of a particular set of modifiers. This class of modifiers is identified as consisting of superlatives, ordinals, and nominal *only*. I propose that for an infinitival clause to receive a non-modal interpretation, it needs to be licensed locally at LF by a modifer from the class identified above. I argue that the appropriate licensing configuration is made available under reconstruction of the head NP into its relative clause-internal position. Recall that the reconstruction option is available to us since we are assuming the analysis of relative clauses proposed in Chapter 2, where the head NP of a relative clause originates in a relative clause-internal position.

The proposed licensing configuration requires that in addition to the relative clause-internal interpretation of the head NP, the licensing element (the superlative *-est*, ordinals, nominal *only* etc.) moves from its NP-internal position to a position where it takes the infinitival clause as an argument. With the licensing configuration in hand, we are able to explain why both nominal *first/last* and relative clause-internal adverbial *first/last* are able to license a non-modal reading. This is so because at LF, which is the point at which the licensing configuration applies, nominal *first/last* and the relative clause-internal adverbial *first/last* are in the same (relative clause-internal) position. We are also able to explain why only superlatives, ordinals, and nominal *only* license non-modal readings and not other nominal modifiers like *many, few, two* etc. These modifiers are able to reconstruct inside the relative clause. However, their semantics is such that they cannot appear in the licensing configuration which requires them to take the infinitival clause as an argument.

The above approach to the licensing of the non-modal reading makes several predictions, all of which are borne out. The first one is that superlative/ordinals/nominal *only*, which associate with focus in general, will not be able to associate with focus in the matrix clause if they are involved in licensing a non-modal interpretation of an infinitival clause. This is so because the licensor has to be interpreted relative clause-internally to license the non-modal interpretation. Association with a matrix focus would require it to move out of the relative clause to a position where it would c-command the focussed element at LF. Relative clauses are islands and therefore such a movement is not possible. We make a further prediction that if the putative licensor does not need to be interpreted internal to the relative clause for licensing reasons, the association with focus readings will be available. This prediction is also borne out.

Further evidence for the possibility of reconstructed interpretations of nominal modifiers comes from the ambiguity of examples like *the first book that John said that Antonia wrote*. These examples permit a 'high' reading (*first* in saying) and a 'low' reading (the x s.t. Jonh said that x was the first book that Antonia wrote). The 'head-raising' analysis proposed in Chapter 2 is required to derive the 'low' reading.

To explore the interpretive properties of the non-modally interpreted infinitival clause, the semantics of the ordinal *first* is used as a probe. A semantics for *first* is proposed and applied to non-modal infinitival relatives. We observe that a temporal simultaneity requirement holds between the head NP and the infinitival clause in a non-modal infinitival relative. The *first 80 yr. old to walk on the moon* must have been 80 yrs. old when he/she walked on the moon. The description cannot pick out the first person who is 80yrs. old and who walked on the moon. The temporal simultaneity facts fall out of the fact that the head NP and the infinitival clause are both interpreted as predicates of times and individuals and that they combine via intersective modification (Predicate Modification) causing their temporal variables to be simultaneous.

We show that there is evidence that the non-modally interpreted infinitival clause comes with temporal/aspectual specification. Two proposals are considered regarding the nature of the temporal/aspectual specification: (i) the infinitival clause contains a covert perfect, (ii) the infinitival clause is specified for perfective aspect if the clause is eventive, and unspecified aspectually if the clause is stative. Asymmetries between infinitival clauses with an overt *have* and those without lead us to adopt the latter proposal.

Next, we consider what conclusions to draw from the fact that in certain cases, a non-modal infinitival relative clause can be interpreted with respect to the future. We show that non-modal infinitival relatives are not in themselves future-oriented. Placed in certain environments, they are evaluated at a future point of time. However, for a non-modal infinitival relative to be interpreted with respect to the future, there must be a future in the neighbourhood of the non-modal infinitival relative which provides a future time at which the non-modal clause can be evaluated.

The chapter concludes with two appendices. The first provides a brief summary of the semantics proposed for *only* and ordinals and the second explores an *in-situ* licensor analysis of non-modal infinitival relatives.

2.3. Chapter 4: Infinitival questions

This chapter addresses two aspects of infinitival questions: their distribution and the interpretation of the modality that appears in infinitival questions. Under distribution, we examine what classes of predicates take infinitival question complements and what generalizations hold of these classes of predicates. With respect to interpretation, we investigate the nature of the modality that appears in infinitival questions and the fact that its force seem to vary depending upon the environment in which it appears.

We begin with a listing of the classes of predicates that take infinitival question complements. It is shown that of the predicate classes in Karttunen (1977)'s list of predicates that take interrogative complements, verbs of conjecture (e.g. *guess*), verbs of relevance (e.g. *matter*), and verbs of dependency (e.g. *depend upon*) do not take infinitival question complements. We observe further that all predicates that take infinitival question complements also take finite question complements. This generalization is explained by proposing that predicates can only s(emantically)-select and not c(ategorially)-select their complement. Since both infinitival questions and finite questions denote questions, any predicate that subcategorizes for the former subcategorizes for the latter. The reverse turns out to not be the case because infinitival questions can only denote a subset of the meanings denoted by finite questions. Hence it is possible for a predicate to subcategorize for finite questions but not infinitival questions. The question of whether predicates that take both infinitival questions and infinitival non-question complements (e.g. *decide*) should be analyzed as involving the same predicate or two distinct but homophonous predicates with distinct semantics is also addressed.

We now turn to the question of the modality that appears in infinitival questions. In Chapter 2, I identified the infinitival [+wh] C^0 as the source of the modality in infinitival questions. I demonstrate that with the exception of a limited set of environments the modality in an infinitival question is always interpreted as deontic/bouletic modality. It cannot be interpreted as circumstantial or epistemic modality.

The force of the modality that appears in infinitival question seems to vary depending upon the environment. A list is provided of the factors that influence the force of the infinitival modality. I argue that despite the seeming variability in the modality of infinitival questions, all infinitival questions involve the same modality. The semantics of the infinitival modality is formalized by the modal operator $\Diamond_{D,\rightarrow}$. The variation between *could* and *should* readings of infinitival questions comes from the contextually salient goals, the nature of the *wh*-phrase, and the semantics of the embedding predicate.

Finally, the proposal made to explain the variability in the force of the modality in infinitival questions is extended to non-subject infinitival relatives. It has been noted by Hackl & Nissenbaum (1999) that the force of the infinitival modality in a non-subject infinitival relative seems to depend upon the determiner that appears with the head of the infinitival relative clause: strong determiners permit only *should* readings, while weak determiners permit both *could* and *should* readings. I show that the mechanism developed for explaining the variation in the force of the modality in infinitival questions extends to provide an explanation of the facts that Hackl & Nissenbaum (1999) observe for infinitival relatives.

2.4. Chapter 5: Ability modals and their actuality entailments

So far, we have examined cases where we interpret a modality that is not overtly present in the structure being interpreted. In this chapter, we will investigate a case which is the mirror image of the cases investigated so far. This is the case of ability modals. Ability modals are analyzed as lexical modals but in certain environments, their presence in a structure is not marked by the presence of any modality in the interpretation.

I propose that the modality of the ability modal is not lexically represented in the lexical item corresponding to the ability modal. It is introduced by a generic modal operator, which occurs in other environments too (e.g. *This machine peels fifty oranges in an hour*). The environments where the ability modal does not introduce any modality into the interpretation are simply environments where the generic operator that introduces modality is absent. The lexical item corresponding to the ability modal itself has the semantics of an implicative verb like *manage*. Crosslinguistic evidence is provided for this proposal by showing that in languages where the perfective/imperfective distinction reflects the absence/presence of genericity, the modal reading is available only in the presence of imperfective aspect. In the presence of perfective aspect, the ability modal behaves like an implicative verb.

Chapter 2
The syntax of infinitival relatives

The purpose of this chapter is to analyze the structure of infinitival relative clauses. We show that while subject infinitival relatives are a kind of reduced relative clause, object infinitival relatives resemble finite relative clauses. A variety of facts about the interpretation of the various kinds of infinitival relatives and infinitival questions will fall out from the syntactic proposals made here.[2]

1. Subject infinitival relatives as reduced relatives

We show that subject infinitival relatives (cf. 3) are reduced relatives[3] and that they combine with their head noun in the manner other reduced relatives (adjectival and participial) combine with their head noun.

(3) a. The [man [___ to fix the sink]] is here.

b. The [book [___ to be read for tomorrow's class]] is kept on the table.

c. The [first man [___ to walk on the moon]] visited my school yesterday.

Some of the characteristics of adjectival and participial reduced relatives are shown in (4).

(4) a. the relativized element is always in the subject position.

b. the subject position does not receive case (from the relative clause).

c. the relativization is very local - only the matrix subject can be relativized.

d. the clausal structure that functions as a reduced relatives can appear as the complement of predicative *be* (cf. Baker (1995), Embick 1997, Iatridou, Anagnostopoulou, & Izvorski 1999).

e. no complementizer is permitted.

f. no relative pronoun is permitted.

These characteristics are shared by subject infinitival relatives. Subject infinitival have characteristic (a) definitionally. Characteristic (b) follows because infinitival clauses do not assign case to their subject position (cf. 4b).

There do not seem to be any instances of subject infinitival relatives formed on non-case marked embedded subjects (cf. 4c). We do not find any case of the following sort.

(5) the N [Op_i [... VP [t_i to VP]]]
(where t_i does not receive case)

The non-existence of examples like (5) cannot be explained by making reference to the absence of case for the [Op_i, t_i] chain since such a reference would also rule out the grammatical examples in (3). The closest that we come to the structure in (5) is with the examples in (6).

(6) a. *the boy [Op_i [for $John_i$ to try [t_i to win]]]
b. *the boy [Op_i [[t_i to dance] would be fun]]
c. *the boy [Op_i [it would be fun [t_i to dance]]]

In general, all embedded non-case marked subjects are controlled: either by an overt argument or by an implicit argument. The Condition C violation caused due to obligatory control is responsible for the ungrammaticality of (6a). (6b) is ruled out by the island violation triggered due to movement out of a subject. It is not clear what rules out (6c). In any case, the fact that there are relativizable non-case-marked matrix subjects but no relativizable non-case-marked embedded subjects is suggestive. We will provide an explanation for this fact.

Infinitival clauses with basically the same interpretation as the subject infinitival relatives in (3a, b) appear as complements to predicative *be* (cf. 4d).

(7) a. The wine is [$_{PurposeClause}$ to complement the cheese].
b. John is [to do his homework before he watches TV]. (Modal *is*)

The construction in (7a) is analyzed and dubbed the presentational *be* construction by Jones (1985a, b) and Jones (1991). The modal *is* construction is discussed by Lees (1960), Pullum and Wilson (1977), and Jones (1985b).[4] The subject infinitival relatives in (3a, b) are modal. The source of the modality in the subject infinitival relatives in (3a, b) is the modality underlying the purpose clause (cf. 7a) and the modality underlying the modal *be* construction (cf. 7b).

Like participial and adjectival reduced relatives, subject infinitival relatives also do not permit the infinitival complementizer *for* to appear.

(8) *the man [Op$_i$ [for [t$_i$ to fix the sink]]]
(compare with *the man [Op$_i$ [that [t$_i$ fixed the sink]]]*, **the man that/for eating the apple*, and **the apple that/for eaten by the man*)

It is possible that (8) is unacceptable due to a violation of the *for*-trace filter. However, the corresponding finite relative clause does not involve a *that*-trace violation despite a similar configuration. This undercuts the plausibility of a *for*-trace account for the ungrammaticality of (8).

Again like participial and adjectival reduced relatives, subject infinitival relatives do not permit any relative pronouns.

(9) *the man [who$_i$ [t$_i$ to fix the sink]]
(compare with *the man [who$_i$ [t$_i$ fixed the sink]]*)

2. Non-subject infinitival relatives

In contrast to what we have argued for subject infinitival relatives, we will argue that non-subject infinitival relatives are not reduced relatives. We will show that non-subject infinitival relatives resemble finite relative clauses and not reduced relative clauses in most respects.

As the name suggests, non-subject positions can be relativized. In fact, case-marked embedded subject positions can also be relativized.

(10) a. a book [Op$_i$ [PRO to read t$_i$]]
b. a knife [Op$_i$ [PRO to cut bread [with t$_i$]]]
c. the person [Op$_i$ [PRO to believe [t$_i$ to be innocent]]]

We will see that relativizations on embedded subjects pattern with non-subject infinitival relatives. It is more accurate to make the cut between matrix subject infinitival relatives and non-matrix-subject infinitival relatives. However, for brevity's sake, henceforth matrix subject infinitival relatives will be referred to as subject infinitival relatives and non-matrix-subject infinitival relatives as object infinitival relatives.

With finite relatives and contra reduced relatives, the relativized position in a non-subject infinitival relative must receive case.

(11) a. *a person [Op_i [PRO to be fond t_i]]

b. a person [Op_i [PRO to be fond of t_i]]

(11a) is ungrammatical because the operator variable chain [Op, t_i] does not receive case. If a case-assigning *of* is present, the example is fine (cf. 11b).

Object infinitival relatives can involve long-distance extraction. We have already seen one example in (10c). (11) presents some more examples.

(12) exs. from Jones (1991), pg. 122

a. Here's [a book [Op_i [PRO to tell your parents [that you're reading t_i]]]].

b. Here's [a book [Op_i [PRO_j to try [PRO_j to get [John to read t_i]]]]].

The fact that object infinitival relatives permit long-distance extraction patterns with finite relative clauses and not with reduced relatives.

The infinitival complementizer *for* can appear freely with object infinitival relatives (as long as the doubly-filled COMP filter is not violated). In this respect, object infinitival relatives resemble finite relative clauses which freely permit the complementizer *that* (again as long as the doubly-filled COMP filter is not violated).

(13) a. a book [Op_i for [you to read t_i]]

b. a knife [Op_i for [you to cut bread with t_i]]

As we saw in (8), reduced relatives do not permit (overt) complementizers.

With finite relative clauses and contra reduced relatives, object infinitival relatives permit relative pronouns (cf. 14a).

(14) a. a knife [[with which]$_i$ C^0 [PRO to cut the bread t_i]]

b. *a knife [[which]$_i$ C^0 [PRO to cut the bread with t_i]]
(compare with *a knife which J. cut the bread with*)

c. *the book [[which]$_i$ C^0 [PRO to read t_i]]
(compare with *the book which J. read*)

d. a knife [Op_i C^0 [PRO to cut the bread with t_i]]

However, unlike finite relative clauses, overt material can be present in the [Spec,CP] of an infinitival relative only if it is part of a pied-piped PP. It is not well understood why this difference exists between finite relative clauses and object infinitival relatives (for proposals see Chomsky & Lasnik (1977), and Pesetsky (1997)).

With the exception of the constraint illustrated in (14), all the properties of object infinitival relatives discussed in the section follow from assuming that object infinitival relatives have a structure that is essentially identical to finite relative clauses i.e. object infinitival relatives are CPs which involve A'-movement of a relative operator/relative pronoun to the [Spec,CP] of the matrix C^0.

Let us quickly go over the relevant properties: (i) non-subject positions can be relativized: Modulo islands and constraints like the ECP, A'-movement can take place from any position; (ii) the relativized position must receive case: the tail of an A'-chain must receive case; (iii) the movement can be long-distance: A'-movement is unbounded as long as conditions on movement like subjacency are not violated; (iv) the infinitival complementizer *for* can appear: Since there is a C^0, it is to be expect that an infinitival complementizer can appear there, (v) overt material can appear in [Spec,CP]: Since object infinitival relatives have a [Spec,CP] position, it is not surprising that relative pronouns can appear there.

3. The modality of infinitival relatives and questions

In this section, I will lay out some of the observations that need to be explained with respect to the availability of modal interpretations for infinitival questions and relatives. I will also make a proposal regarding the modality of infinitival questions and object infinitival relatives.

The first observation is that all instances of infinitival questions and object infinitival relatives are interpreted as modal (cf. Kjellmer 1975, Geisler 1995, Pesetsky and Torrego 2001).

(15) a. Hafdis knows [who_i C^0 [PRO to talk t_i to at the party]].

b. The people [Op_i C^0 [PRO to talk t_i to at the party]] are Magnus, Herb, and Penna.

The second is that subject infinitival relatives may be modal or non-modal (cf. Kjellmer 1975, Geisler 1995, Pesetsky and Torrego 2001).

(16) a. The [man [___ to fix the sink]] is here.

b. The [book [___ to be read for tomorrow's class]] is kept on the table.

c. The [first man [___ to walk on the moon]] visited my school yesterday.

(16a, b) only have modal interpretations. In contrast, (16c) has a non-modal interpretation. The exact conditions under which a non-modal interpretation is available and the interpretation of non-modal infinitival relatives like (16c) is the subject of chapter 3.

The third and final observation relevant here is that the modality in modal subject infinitival relatives is distinct from the modality in object infinitival relatives and infinitival questions. The force of the modality in modal subject infinitival relatives seems to be independent of its environment in a way that the force of the modality in object infinitival relatives and infinitival questions is not (cf. 17).

(17) a. Subject infinitival relative ('purpose clause')
  i. A man to fix the sink ($\approx$ a man whose purpose is to fix the sink)
  ii. The man to fix the sink ($\approx$ the man whose purpose is to fix the sink)

b. Subject infinitival relative ('modal *is*')
  i. A book to be read ($\approx$ a book which is to be read)
  ii. The book to be read ($\approx$ the book which is to be read)

c. Object infinitival relative (cf. Hackl & Nissenbaum 1999)
  i. a book [Op_i C^0 [PRO to read t_i]] ($\approx$ a book that one could/should read)
  ii. the book [Op_i C^0 [PRO to read t_i]] ($\approx$ the book that one should read)

d. Infinitival question
  i. Hafdis knows [where_i C^0 [PRO to get gas t_i]]
  ($\approx$ Hafdis knows where she can get gas)
  ii. Hafdis knows [who_i C^0 [PRO to talk to t_i at the party]]
  ($\approx$ Hafdis knows who she should talk to at the party)

We see that while the force of the modality seems to sometimes be existential and sometimes universal with object infinitival relatives and infinitival questions, there is no such variation with modal subject infinitival relatives. This suggests that the modalities in modal subject infinitival relatives and in object infinitival relatives and infinitival questions are likely to be distinct, and that it is likely that the modality in object infinitival relatives and infinitival questions are closely related/identical (see Chap. 4, sect. 4).

3.1. Object infinitival relatives and infinitival questions

The source of the modality in object infinitival relatives and infinitival questions is the [+wh] infinitival complementizer, using [+wh] to generalize over the relative and the interrogative complementizer.[5]

(18) C^0[+rel/wh,+inf] is interpreted as the modal $\Diamond_{D,\rightarrow}$

The semantics of $\Diamond_{D,\rightarrow}$ will be provided in Chapter 4. (18) abstracts away from the additional semantics that C^0[+wh,+inf] has by virtue of being an interrogative complementizer.

All instances of infinitival questions and object infinitival relatives are modal because their structure involves the infinitival [+wh] complementizer which is interpreted as a modal. Since they share the same modality, the existence of variable force effects (cf. 17c, d) in both is not suprising. The variable force effects will fall out of the semantics provided for $\Diamond_{D,\rightarrow}$ in Chapter 4.

One prediction made by this syntactic proposal is that the modality in object infinitival relatives and infinitival questions is very high in the structure. Hence, we do not expect to see it getting negated by sentential negation. This claim is borne out (see Chap. 4, sect. 3.5).

3.2. Subject infinitival relatives

We have noted that subject infinitival relatives can be modal (cf. 19a, b) as well as non-modal (cf. 19c).

(19) a. The [man [___ to fix the sink]] is here.

b. The [book [___ to be read for tomorrow's class]] is kept on the table.

c. The [first man [___ to walk on the moon]] visited my school yesterday.

Under our proposal, subject infinitival relatives are reduced relatives and reduced relatives, we will argue do not involve a C^0 projection. This is good because if all subject infinitival relatives involved a C^0 projection and if, as has been proposed above, infinitival C^0 receives a modal interpretation, we would be hard put to explain why (19c) can receive a non-modal interpretation. In addition, we would be unable to explain why the modally interpreted subject infinitival relatives in (19a, b) do not display variable force effects (cf. 17).

The proposal is that the source of the modality in (19a, b) lies in the infinitival. The relationship between (20a) and (20b), and the relationship between (21a) and (21b), is the same as the relationship between (22a) and (22b).

(20) a. The wine is [$_{XP}$ to complement the cheese].

b. The [wine [$_{XP}$ to complement the cheese]] is in the cellar.

(21) a. The book is [$_{XP}$ to be read for the class tomorrow].

b. The [book [$_{XP}$ to be read for the class tomorrow]] is kept on the lectern.

(22) a. The book is [$_{PP}$ on the table].

b. The [book [$_{PP}$ on the table]] is delightful.

(20a) and (21a) are modal for whatever reason (20b) and (21b) are modal. Unlike infinitival questions and non-subject infinitival relatives, the question formation/relativization is not responsible for the modality.

We will discuss the case of the non-modal (19c) at some length in Chap. 3. To preview, (19c) receives a non-modal interpretation because it contains a non-modally interpreted infinitival clause. This non-modally interpreted clause can only occur if 'licensed' by a particular class of elements in a particular domain.

3.3. Crosslinguistic implications: languages without infinitives

The discussion in this section is largely based on comments in Giannakidou (2003b)'s review of the dissertation that this book is based on. Giannakidou notes that in Greek, which is a language without infinitival clauses, clauses marked with the particle *na* carry much of the functional load of infinitives in English.[6] For example interrogative *na* clauses provide the equivalent of English infinitival questions.

(23) Greek *na* questions:

I Ariadne kseri/anarotiete [[me pjon] na milisi].
the Ariadne knows/is.wondering with whom NA talk.3Sg

'Ariadne knows/wonders who to talk to.'

But such clauses cannot be used as relative clauses, subject or object. The presence of the purpose complementizer does not help.

(24) Greek *na* relatives:

a. subject *na* relative: *

**O anthropos [(ja) na ftiakisi to neroxiti]*
the man for NA fix.3Sg the sink

b. object *na* relative: *

**To vivlio [(ja) na diavasis]*
the book for NA read.2Sg

The presence of elements that would license a non-modal reading for a subject infinitival relative clause in English does not help either.

(25) Greek subject *na* relative with 'first': *

**O protos anthropos [na perpatisi sto fegari]*
the first man NA walk.3Sg on-the moon

In our analysis of English subject infinitival relatives we have related cases like *The wine to complement the cheese* to cases like *The wine is [to complement the cheese].* The corresponding Greek data suggests that additional variable are active in this domain. As we have seen above, Greek does not permit any kind of *na*-relative. But it does permit the counterpart of *The wine is [to complement the cheese].*

(26) Greek

a. *To krasi ine [ja na sinodepsi to fajito].*
the wine is for NA complement.3Sg the meal
'The wine is to complement the meal.'

b. **To krasi [ja na sinodepsi to fajito].*
the wine for NA complement.3Sg the meal

We cannot do full justice to the Greek data here but the following lines of inquiry suggest themselves. The first pursues the intuition that only phrases that are sufficiently nominal can function as reduced relatives. According to this line of inquiry , *na* clauses would not be sufficiently nominal. Given the presence of person agreement in *na* clauses, this claim seems reasonable. The second line of inquiry would explore the crosslinguistic status of the connection between an XP functioning as a complement of *be* and the same XP functioning as a reduced relative. In many languages, PPs can function predicatively with *be* but not as a reduced relative. This is the case in Hindi-Urdu.

(27) Hindi-Urdu:

a. Predicative PP:

billi kamre-mẽ hE
cat.f room-in be.Prs.Sg
'The cat is in the room.'

b. PP as a reduced relative: *

**[[kamre-mẽ] billi] moṭii hE*
room-in cat.f fatf .be.Prs.Sg
Intended: 'The cat in the room is fat.'

If there is no crosslinguistic entailment from an XP functioning as a complement of *be* to the XP functioning as a reduced relative, then (26) poses less of a puzzle.

Finally, the question remains why *na* clauses can function as infinitival questions but not as infiniitval relatives, full or reduced. Giannakidou (1998)'s analysis of *na*-clauses as subjunctive might be helpful here. If subjunctive clauses need licensing, their distribution might fall out from their licensing conditions. *na* questions would be locally licensed by their embedding predicate. Such a licensing configuration would not be available to *na* relative clauses. These remarks are, of course, quite tentative and meant to inspire further work on the distribution of non-finite clause types. A proper analysis would also need to consider the fact that subjunctive relative clauses are well documented in Romance (cf. Farkas (1998)) and contrast their licensing conditions with the licensing conditions of *na*-clauses in their various incarnations: purpose clauses and questions but not relative clauses.

4. Structures for reduced relatives

Having decided on a structure for object infinitival relatives, we will now consider structures for subject infinitival relatives. In sect. 1, it was shown that subject infinitival relatives pattern with reduced relatives. Therefore, we will consider some structures for reduced relative clauses. We will start with a discussion of a proposal made by Kayne (1994) for reduced relative clauses. In sect. 5, a summary of the arguments for a Raising analysis of relative clauses will be provided. Finally, in sect. 6, we will propose a structure that resembles Kayne's analysis in spirit but not in execution.

4.1. Kayne (1994)'s proposal for reduced relatives

4.1.1. Kayne (1994)'s general proposal for relative clauses

Kayne (1994) proposes that all relative clauses involve a head-raising analysis. In his proposal, relatives clauses are CP complements of D^0. There is no CP-external NP head of the relative clause.

(28) a. $[_{DP}$ $[_{D^o}$the] $[_{CP}$ book$_i$ $[_{C'}$ that [Gretchen read t_i]]]]

b. $[_{DP}$ $[_{D^o}$the] $[_{CP}$ book$_i$ $[_{C'}$ C^0 [Gretchen read t_i]]]]

c. $[_{DP}$ $[_{D^o}$the] $[_{CP}$ $[_{DP}$ book$_i$ [which t_i]]$_j$ $[_{C'}$ C^0 [Gretchen read t_j]]]]

In (28), the 'head' NP and the relative pronoun together form a constituent. On the other hand, the relative pronoun and the *C′* do not form a constituent. It is however possible to co-ordinate two sequences of a relative pronoun and its associated *C′* (cf. 29a) but it is not possible to co-ordinate two sequences of an NP and its associated relative pronoun (cf. 29b).

(29) a. the book [[which$_i$ Gretchen likes t_i] and [which$_j$ Petra dislikes t_j]]

b. *the [[[book which] and [cartoon which]]$_i$ Gretchen likes t_i] (compare with 'Which books and what music does Gretchen like?')

Kayne provides some arguments that show that a head-raising analysis must be available. These and some other arguments for a head-raising analysis are reviewed in the next section (sect. 5). He does not provide any arguments that show that the commonly assumed head external analysis of relative clauses which involves right adjunction of CP to NP is not feasible. Since he is working within a system which does not permit right-adjunction, the right adjunction analysis of relative clauses is never a potential candidate. If we assume that in addition to the head-raising analysis, the head-external analysis is also available, we can explain why co-ordination is possible in (29a). It is possible because under the head-external analysis, the relative pronoun and the *C′* associated with it form a constituent. This solution works but it has an undesirable prediction. Since (29a) only has a head-external analysis, we would expect it not to be possible to interpret the head of the relative clause inside the relative clause. (30) is a counterexample to this prediction.

(30) the picture of his$_i$ [[which every freshman$_i$ likes] and [which every sophomore$_i$ dislikes]]

Further, the ungrammaticality of (29b) is still a puzzle.

4.1.2. *Kayne (1994)'s proposal for reduced relatives*

Kayne proposes that reduced relatives and finite relative clauses have essentially the same structure. They differ in the kind of complement the C^0 takes. In case of a finite relative clause, C^0 takes a finite IP complement and in the case of a reduced relative, C^0 takes a participial/ adjectival complement. In the following discussion, I will refer to this complement by the cover term PrtP (participial phrase). The use of this cover term is meant to extend to adjectival and prepositional phrases and is not meant to imply that the different kinds of reduced relative clauses are based on exactly the same functional projection.

(31) a. the [$_{CP}$ book$_i$ [C^0 [$_{PrtP}$ t$_i$ sent to me]]]

b. *the [$_{CP}$ [book$_i$ [which t$_i$]]$_j$ [C^0 [$_{PrtP}$ t$_j$ sent to me]]]

He proposes that *book* in (31a) receives case via (LF-)incorporation into *the*. This case, he proposes is sufficient to case-license the trace of *book* in [Spec,PrtP]. (31b) is ruled out by noting that the case acquired by *book* is not sufficient to case-license the trace of *which book* in [Spec,PrtP].

Kayne also rules out (32a) by appealing to a failure of case-licensing in the [Spec,IP].

(32) a. *the [$_{CP}$ book$_i$ [C^0 [$_{PrtP}$ Jonah fond of t$_i$]]]

b. *the [$_{CP}$ book$_i$ [C^0 [$_{PrtP}$ PRO fond of t$_i$]]]

No explanation is provided for the ungrammaticality of (32b). A suggestion is made in footnote 33 about the ungrammaticality of (32b) following from conditions on the licensing of null case. Probably the participial IP is unable to license null case to the PRO. However, this assumption is problematic since we have cases like (33) which plausibly involve a participle with a PRO subject.[7]

(33) [PRO$_i$ humming a song], Mary$_i$ went to school.

Within Kayne's system, we can explain the ungrammaticality of examples like (32b) by making the following assumption: the subject position of a reduced relative does not receive any kind of case, Null or other. By the extended projection principle, this position needs to be projected and filled. Therefore whatever is generated in this position must get case from elsewhere.

Kayne assumes that the 'head' of the relative clause is able to get case via (LF-)incorporation into D^0. I will refer to this mechanism of case-licensing as **Case Transmission**. In (32b), *book* gets case but the PRO doesn't, leading to ungrammaticality.

We now have an explanation for a generalization about reduced relatives: only subjects can be relativized. This follows from the assumption that reduced relatives never license case on their subject position. So the only way for the subject of a reduced relative to be case-licensed is by being relativized and receiving case via the mechanism of case transmission. Therefore if the subject of a reduced relative is not relativized, it will not be case-licensed and the structure will be ungrammatical.

The initial conceptual problem that we started with was the putative existence of A'-chains that do not receive case, which would be the case if there is relativization from the subject position of a reduced relative. With the mechanism of case-transmission proposed by Kayne, the movement chain created by relativization in a reduced relative does receive case and hence reduced relatives cease to be conceptually problematic. However, the mechanism of case-transmission is empirically problematic. The only place where it seems to do any work is with reduced relatives.

(34) a. The [book$_i$ C^0 [Miguel likes t$_i$]] is on sale.

b. * The [book$_i$ C^0 [Miguel is fond t$_i$]] is on sale.

c. The [book$_i$ C^0 [Miguel is fond of t$_i$]] is on sale.

In (34a), *book* ends up receiving both accusative and nominative case. This is perhaps not troubling for English since English does not display (for the most part) case-matching effects. However, one would expect the counterparts of (34a) to be ungrammatical in languages which do display case-matching effects e.g. German. This turns out not to be the case - languages that display case-matching effects display them in free relatives and not in headed relatives.

(34b) raises a more severe problem. By case-transmission, *book* in (34b) should receive nominative case and the sentence should not be ungrammatical. That it is ungrammatical for case reasons is demonstrated by the grammaticality of the minimally different (34c).[8]

I conclude therefore that Kayne's analysis of reduced relatives is problematic. The explanation depends upon the empirically problematic mechanism of case-transmission.

Since Kayne's analysis is a head-raising analysis, it allows reconstruction into the relative clause/ interpretation of the material in the head of the relative clause inside the relative clause. I believe that the arguments for a head-raising analysis survive the problems with Kayne's particular proposal and the analysis that I will propose will involve an analysis of relative clauses (finite and reduced) that permits head-raising.

5. Arguments for a raising analysis of relative clauses

5.1. The candidates

Syntactic analyses of relative clauses within the generative literature can be grouped into three classes: the head external analysis, the matching analysis, and the raising/promotion analysis. These three approaches are schematically presented in (35).[9]

(35) the book which Mona likes

a. Head-External Analysis[10]
[the [$_{NP}$ book] [$_{CP}$ which$_i$ C^0 [Mona likes t$_i$]]]

b. Matching Analysis (Lees 1960, 1961, Chomsky 1965)
 [the [$_{NP}$ book] [$_{CP}$ [which *book*]$_i$ C^0 [Mona likes t$_i$]]]
 (italicized *book* is deleted at PF under identity)

c. Raising(/Promotion) Analysis (Schachter 1973, Vergnaud 1974)
 [the [$_{NP}$ book]$_j$ [$_{CP}$ [which t$_j$]$_i$ C^0 [Mona likes t$_i$]]]

In the matching analysis, the head NP does not originate inside the relative clause. Under the raising analysis, the head NP actually originates inside the relative clause and moves out of the relative clause to a relative clause external NP position. A reflex of this difference between the two analyses is that while the Raising/Promotion analysis permits the head NP to be interpreted *wholly* inside the relative clause, the matching analysis requires the head NP to be interpreted outside the relative clause. 'Full Reconstruction' into the relative clause is an option only under the Raising/Promotion analysis since only in the Raising/Promotion analysis does the head NP originate inside the relative clause.

The three analyses in (35) can be seen as lying on a continuum with the head-external analysis at one end, the raising analysis at the other, and the matching analysis in between. Under the head-external analysis, the information about the head NP is not syntactically available inside the relative clause. There is no syntactic relationship between the head NP and the relative operator chain. The matching analysis makes information about the head NP syntactically available inside the relative clauses. There is, however, no syntactic relationship between the head NP and the relative operator chain. Finally, the raising analysis makes information about the head NP available inside the relative clause and the head NP originates inside the relative clause. Under the raising analysis, depending upon where the head NP is interpreted at LF (i.e. which copies of the head NP are interpreted), we get the effect of a 'head-external' analysis or a 'head-internal' analysis.

The arguments that follow show that in the particular cases under discussion, the head NP must be interpreted internal to the relative clause i.e. in these cases, the head/external matching analysis is not feasible; only the head raising analysis is available. They do not show that the 'head-external' /'matching' analysis is ruled out in other cases. Arguments that rule out a head-external analysis across the board will have to be of a broader conceptual nature and will be related to the question of why certain predicates can function as nominal modifiers (e.g. Participial Phrases, PPs, APs) but not others (e.g. VPs).[11]

5.2. The argument from 'Idioms'

The first argument is from the domain of "idiom" interpretation/ interpretation of fixed expressions. It works as an argument against the "matching" analysis and the head-external approach.

The logic behind the argument from "idioms" for a raising analysis goes as follows: the "idiom" can only appear as part of a larger expression. The larger expression is shown in the (a) examples. That it cannot appear outside this context is shown in (b). However, the "idiom" is able to felicitously appear as the head NP of a relative clause, where the position it is associated with inside the relative clause (its trace) is part of the larger expression it needs to appear with. This is shown by the examples in (c).

Under a "matching"/"head-external" analysis, given the unacceptability of the examples in (b), the acceptability of the examples in (c) is unexpected and vice versa. The raising analysis is able to explain these facts parsimoniously. "Idioms" need to appear in a particular environments as shown by the examples in (a) and the unacceptability of the examples in (b). The examples in (c) are acceptable because the "idiom" appears in the relevant environment at some point in the derivation (minimally point of Merge, maybe also at LF).

(36) (cited to Brame 1968 ms., ex. 35 from Schachter 1973)

a. We made headway.

b. * (The) headway was satisfactory.

c. The headway that we made was satisfactory.

(37) (ex. 36 from Schachter 1973)

a. She's keeping careful track of her expenses.

b. * (The) careful track pleases me.

c. The careful track that she's keeping of her expenses pleases me.

(38) (ex. 37 from Schachter 1973)

a. Lip service was paid to civil liberties at the trial.

b. * I was offended at (the) lip service.

c. I was offended by the lip service that was paid to civil liberties at the trial.

(39) (cited to George Bedell, fn. 15 from Schachter 1973)
 a. He solved the problem in a clever way.
 b. The clever way in which he solved the problem impressed me.
 c. *The clever way impressed me.

5.2.1. A related argument from subcategorization

Larson (1985) observes that headed relative clauses containing a trace in adjunct position, but neither a relative adverb or a stranded preposition, are grammatical only if the head NP is a *bare-NP adverb*.[12]

(40) a. the way [Op_i that you talk t_i]
 b. *the manner/fashion [Op_i that you talk t_i]
 c. You talk that way.
 d. *You talk that manner/fashion.

The well-formedness of the operator-variable chain in (40a) depends upon what the head NP is. Information about the head NP is required internal to the relative clause. Under a head-raising analysis, the ill-formedness of (40b) directly follows from the ungrammaticality of (40d). This explanation is not directly available under the head-external analysis and Larson has to introduce a feature-transmission mechanism which makes the relevant information about the head NP available internal to the relative clause.

5.3. The argument from binding theory

The second argument comes from the domain of binding. This argument was originally provided by Schachter (1973) as an argument against the "matching" analysis. It works as an argument against the head-external analysis too. The argument is based on the examples in (41-43), which show that here for the purposes of binding theory the head NP behaves as if it was in its (lowest) trace position inside the relative clause.

Under a "head-external" analysis, we expect the Condn. A to be bad (unless we appeal to a device such as logophoricity etc.) and we expect the Condn. B and Condn. C examples to be good. The matching analysis would also have trouble with the Condn. A facts.

The actually observed pattern of (un)grammaticality can be explained under the raising analysis if the "head NP" is interpreted at LF in its trace position. All the binding theory examples below are cases where like in the "idiom" cases (*paint a portrait of, show an interest in, have a (Adj) opinion of*) it is plausible that at LF the "head NP" has to form a unit at LF with relative-clause internal material. This might explain why we get reconstruction of the head NP into the relative clause for binding purposes.[13]

Condn. A effects shows that the head NP *can* be interpreted internal to the relative clause.

(41) (exs. 42a, 43a from Schachter 1973)

a. The portrait of himself$_i$ that John$_i$ painted is extremely flattering.

b. The interest in each other$_i$ that John and Mary$_i$ showed was fleeting.

Condn. B and C (42 and 43 respectively) show that in the cases at hand, the head NP *must* be interpreted relative clause internally.

(42) *The opinion of him$_i$ that John$_i$ has is favorable.
(compare with: *The opinion of himself$_i$ that John$_i$ has is favorable.*)

(43) (exs. 41b, 42b from Schachter 1973)

a. *The opinion of John$_i$ that he$_i$ thinks Mary has is unfavorable.

b. *The portrait of John$_i$ that he$_i$ painted is extremely unflattering.

5.4. Amount relatives

The existence of amount readings provides another reason for assuming a matching/raising analysis of relative clauses. In (44), for independent reasons, we do not wish to entertain a variable following *there be* (cf. Carlson 1977, Heim 1987).

(44) (from Heim 1987:33)
The very few books that there were on his shelves were all mysteries.

LF: The very few λd that there were d-many-books on his shelves were all mysteries.

Therefore it is postulated that the head NP is reconstructed in the trace position and the abstraction is over a degree variable. It is possible to interpret the head NP in the trace NP only under the matching/raising analyses. Thus the analysis of amount relatives provides independent support to the matching/raising analysis.

It should be noted that a matching/raising analysis is required whenever we have an amount reading and not just when we have an extraction from a position subject to the definiteness constraint. This can be seen in (45), which permits a reading where an identity of amounts (without the additional identity of substance) suffices for truth.

(45) (from Heim 1987:40)
It will take us the rest of our lives to drink the champagne that they spilled that evening.
LF1: the λd [that they spilled d-much champagne that evening] (*identity of amounts*)
LF2: the champagne λx [that they spilled x that evening] (*identity of substances*)

LF2 shows the reading we would get if we did not interpret the head NP internal to the relative clause and do a degree abstraction. LF2 is the LF yielded by the head-external analysis and it gives us the wrong, identity of substances, reading. In order to get the degree abstraction, it is necessary to interpret the head NP inside the relative clause i.e. amount readings require the matching/raising analysis.

Reconstruction in amount readings can take the head NP below another scope bearing element thus producing scope reconstruction effects. This is the the case in (46).

(46) (exs. from Sauerland 1998:54a, b)

a. No linguist would read the many books Gina will need for vet school.
need > many

b. Mary shouldn't even have the few drinks that she can take.
can > few

The relatives clauses in (46) have an amount relative reading where the head NP is interpreted under the scope of a modal inside the relative clause. In (46a), for example, there is a cardinality d, that Gina will need d-many books for vet school. No linguist would read *d*-many books. The relevant LF is shown in (47).

(47) No linguist would read [the λd [Gina will need *d*-many books for vet school]]

The relative clause in the LF, however, only returns a degree. We need to extract information about what this is a degree of[14]. Assuming that that can be done, we have the desired reading.

There is no obvious way under a head-external analysis to reconstruct *many books* under the scope of the modal *need* in (46). Under a raising analysis, we have the structures that we need to generate the appropriate LFs.

5.5. Scope reconstruction

In (46), we saw an instance of the head NP taking scope under a relative clause internal modal. (46), however, involved amount readings. The head NP can take scope under a relative clause internal operator even in the absence of amount readings. This point is made by examples like (48).

(48) I am worried about the twenty five people likely to come for dinner tomorrow.
likely $>$ 25 people

In (48), there is the added question of exactly what kind of object the NP + relative clause denotes and how this object is derived from the relative clause. Setting aside that question, it is clear that there is a reading where the worry is about a proposition like 'It is likely that 25 people will come for dinner tomorrow' i.e. the head of the relative clause is interpreted under the scope of *likely* i.e. inside the relative clause. Under the head-external analysis, there would be no way to interpret the head NP under a relative clause-internal operator. The raising analysis gives us the right structure to generate the scope-reconstruction reading though, of course, more needs to be said about how we arrive at the right interpretation.

5.6. Lower readings of adjectival modifiers

The adjectival modifiers in (49) have a 'high' and a 'low' reading.

(49) The first book that John said that Tolstoy had written

'high' reading ≈ In 1990, John said that Tolstoy had written *Anna Karenina*; in 1991, John said that Tolstoy had written *War and Peace*. Hence the NP in (49) is *Anna Karenina*
(i.e. order of *saying* matters, order of *writing* is irrelevant)

'low' reading ≈ John said that *War and Peace* is the first book that Tolstoy had written. Hence the NP in (49) is *War and Peace*.
(i.e. order of *writing* matters, order of *saying* is irrelevant)

Like (49), (50a, b) also have 'high' and 'low' readings.

(50) a. The only book that John said that Tolstoy had written

b. The longest book that John said that Tolstoy had written

The existence of the 'low' readings is puzzling under a head-external analysis of Relatives Clauses, under which the head NP and the relative clause are both predicates which combine to create a new predicate. *first/only/est* apply to this predicate. So we get the first/only/longest member of the set of books s.t. John said that Tolstoy wrote them. This is the 'high' reading. There seems to be no way to put *first/only/est* in the scope of *say*, which is what the 'low' reading requires.

The ambiguity of (49, 50) can be derived using the raising analysis. The raising analysis allows for the head NP to be interpreted inside the RC. For the 'low' readings of (49, 50), the head NP would be interpreted in the embedded clause.

(51) The [John said that [Tolstoy had written first/only/longest book]]

For interpretive reasons, *first/only/est* moves from its surface position to a position where it takes a propositional argument (Szabolcsi 1986, Heim 1995). This yields (52).

(52) The λx [John said that [first [[book,x][that Tolstoy had written x]]]

More needs to be said about exactly how the LF in (52) is interpreted. However, it clear that in (52), *first/only/est* are in the scope of *say*, which is what we need to show in order to argue for the head raising analysis. The details of the analysis can be found in Bhatt (2002).

6. Our proposal for reduced relatives

6.1. Version 1: Direct Predication

The proposal is that the reduced relative clause is a projection smaller than a CP (call it PrtP) that combines with the head NP via intersective modification. The PrtP has a PRO subject and we shall show that it is of the right type to form a predicate that can compose with the noun phrase i.e. $< e,t >$.[15] No movement operation is required to create this predicate.

We assume that all A'-chains need case. The absence of case is what is responsible for * *the book Miguel is fond*. The case requirement of A'-chains makes it impossible to relativize a non-case-marked position by A'-movement. In reduced relatives, we relativize on the subject position of the PrtP which does not receive case (setting aside the issue of Null Case for PRO). This relativization on a non-case-marked position is not ruled out since it does not involve A'-movement. No A'-movement is necessary for relativization (i.e. predicate formation) on the subject position because, as we will show, a PrtP with a PRO subject is a predicate and hence of the right semantic type to combine with the head NP by Predicate Modification.

A PrtP with a non-PRO subject denotes a proposition. I will call the mechanism by which a PrtP with a PRO subject comes to be interpreted as a predicate *Direct Predication*. The nature of Direct Predication is made explicit in sect. 6.2. Unlike A'-movement which is unbounded, Direct Predication is strictly local. We can only relativize on the top-most/outermost unsaturated argument of the predicate.

At this point, it should be noted that we are giving up on assimilating the operation that creates reduced relatives (= Direct Predication) with the operation that creates finite relative clauses (= A'-movement). We believe that such a move is justified because the two operations have distinct properties.

In sect. 4.1, it was shown how Kayne's assimilatory analysis was problematic. Another attempt to provide a unified analysis for finite relative clauses and reduced relative clauses utilized the device of *whiz* (= *wh* + be) deletion. Convincing arguments against the *whiz*-deletion approach are provided in Huddleston (1971), Hudson (1973), Berman (1973), and Williams (1975).

The properties of reduced relatives in general, and subject infinitival relatives in particular, follow from our proposal. Let us go over these properties one by one. Firstly, only subjects can be relativized in a reduced relative:[16]

Reduced relatives lack a C^0 which would be needed for A'-movement which would be necessary for non-subject relativization. The only strategy available for creating a predicate out of the participial clause is Direct Predication which only permits relativization of the subject.[17]

Moreover, the relativized argument can only be the matrix subject: Relativization of a non-matrix subject would require A'-movement of a null-operator, for which we would need a C^0. In the current proposal, reduced relatives have no CP projection and hence non-matrix subjects cannot be relativized.

Finally there are never any relative pronouns/complementizers in a reduced relative: relative pronouns occupy the [Spec,CP] and complementizers occupy C^0 - since reduced relatives lack a CP layer, the absence of relative pronouns and complementizers is to be expected.

6.2. On the nature of Direct Predication

In this section, I will compare the mechanism of direct predication with another mechanism that we can use for predicate formation in Reduced Relatives. I will call this latter mechanism 'PRO A'-movement'. It will be shown that while both mechanisms work, the Direct Predication account is to be preferred since it requires fewer stipulations.

The direct predication mechanism is exemplified in (53). Direct predication assumes that the PRO subject is semantically vacuous. Interpreting the structure '[PRO available]' yields us an object of semantic type <e,t>. It is therefore of the right semantic type to combine with the NP via Predicate Modification. The PRO is not required to A'-move for the purposes of predicate formation.

(53) the [[$_{NP}$ stocks] [$_{AP}$ PRO available]]
$[\![[_{NP} stocks]]\!] = \lambda x.stocks(x)$
$[\![[_{AP} PRO\ available]]\!] = \lambda x.avail(x)$
$[\![the\ stocks\ [PRO\ available]]\!] = \iota(\lambda x.[stocks(x) \wedge avail(x)])$

The PRO A'-movement mechanism is exemplified in (54). A PRO subject is projected. The PRO subject then A'-moves to create a predicate. It is assumed to be semantically vacuous; its effect is taken to be limited to creating a predicate in a manner similar to the treatment of relative pronouns in Heim & Kratzer (1998). Since PRO does not require case, it is assumed that the

A'-chain involving a PRO also does not require case.

(54) the [[$_{NP}$ stocks] [$_{AP}$ PRO [$_{AP}$ t$_{PRO}$ available]]]
$[\![[_{AP}PRO\ [_{AP}t_{PRO}\ avail]]]\!] = \lambda x.avail(x)$
(the AP is generated with a PRO subject, which then moves like a relative pronoun to form a predicate and disappears.)
$[\![the\ stocks\ available]\!] = \iota(\lambda x.[stocks(x) \wedge avail(x)])$

In the case at hand, the two mechanisms give the same results.

The two approaches differ in whether the PRO A'-moves for predicate formation. Both the approaches need to assume A-movement of PRO to handle examples like (55). The PRO subject of *likely* moves there by A-movement from the subject position of the complement clause of *likely*.

(55) The student likely to win the race.
the [[student] [PRO λx likely [x to win the race]]]
(A-movement of PRO for EPP reasons; this step is shared by the *Direct Predication* as well as the *PRO A'-movement* approach.)

Direct Predication:
the [[student] [λx [likely [x to win the race]]]]
(the role of PRO is purely syntactic, it is semantically vacuous)

One-step PRO movement:
the [[$_{NP}$ stocks] [$_{AP}$ PRO [$_{AP}$ t$_{PRO}$ λx [likely [x to win the race]]]]]
(A'-movement of the PRO creates a predicate, the moved PRO is semantically vacuous)

Once again, both approaches yield identical results. However, the locality of the operation involved in reduced relative formation follows directly from the mechanism of Direct Predication but not from the PRO A'-movement mechanism. We need to stipulate that the A'-movement involved in the PRO A'-movement mechanism is always very local. The PRO can only move from the matrix subject position. It cannot cross any heads. Therefore, we will continue to use the mechanism of Direct Predication in our analysis of reduced relatives.[18]

6.3. Version 2: accommodating reconstruction

The proposal as it stands seems to predict that there should be no reconstruction into relative clauses. This is so because we have not postulated any connection between the head NP and a relative clause-internal location.

(56) the [[$_{NP}$philosopher] [λx [x reading the Dworkin book]]] is obtuse.

However, there are reasons to believe that there is reconstruction into reduced relatives.

(57) a. Idiom interpretation:
  i. The headway [made] was considerable.
  ii. The pictures [taken by Sally Harding] inspired many artists.

 b. Scope Reconstruction:
  i. I am worried about the twenty five people likely to come for dinner.
  ii. The twenty five people likely to come for dinner might be a problem.

In general, the arguments that show that the head NP of a finite relative clause can be reconstructed into the relative clause (see sect. 5) carry over to reduced relatives also. These reconstruction effects follow naturally within a Raising analysis of relative clauses. [19] An analysis of finite and reduced relative clauses that allows for reconstruction is provided in (58).

(58) a. Finite relative clauses, object infinitival relatives
 D^0 [*NP* [$_{CP}$ Rel-*NP* C^0 [......Rel-*NP*......]]]
 the [**book** [$_{CP}$ [which **book**]$_j$ C^0 [Jonah likes [which *book*]$_j$]]]

 b. Reduced relative clauses
 D^0 [*NP* [$_{PrtP}$ *NP* [$_{Prt'}$ ....]]]
 the [**philosopher** [$_{PrtP}$ **philosopher** reading Dworkin]]

Relative clauses in this proposal have a CP/PrtP-external NP head. In the case of finite relative clauses, the CP-external head is identical to the complement of the relative pronoun and in the case of reduced relatives, the PrtP external head is identical to the subject of the reduced relative clause. (We address in the next paragraph how the identity between the bold-faced NPs in (58) is enforced.) Since the head NP is represented inside the relative clause, it becomes possible to derive reconstruction effects. The syntactic material corresponding to the head NP can be interpreted in various positions in the chain to which the relative phrase in the [Spec,CP] of the finite relative clause/the subject of the reduced relative belongs. Depending upon which copy of the chain, the syntactic material corresponding to the head NP is interpreted, we get the effects of a 'head external' and a 'head raising' analysis. If we interpret the highest copy, we get the 'head-external' analysis and if we interpret a lower copy, we get a 'head-raising' analysis.

Let us consider two different ways in which the identity relationship between the head NP and the CP/PrtP-internal NP can be enforced. The first way - the matching analysis - does not assume that the head NP and its relative clause-internal representation are related by movement. We stipulate an LF-condition which forces the head NP and the complement of the relative operator in the finite relative clause and the subject in the reduced relative to be identical. In addition, we would also need to stipulate that the relative clause-internal material identical to the head NP is obligatorily deleted at PF. This approach is the matching analysis applied to reduced relatives and as such inherits the limitations of the matching analysis - it will work only in the environments where the matching analysis works.

The other way, which is adopted here, relates the head NP and the relative clause internal material by overt syntactic movement. If all instances of relative clauses involved the raising analysis, this movement would be an unorthodox one in that we would need to permit extractions of unboundedly deeply embedded possessors. See Bhatt (2002) for an argument that cases that involve extraction of possessors do not permit the raising analysis; their derivation involves the matching analysis.

(59) Assumption: *which* with a covert [+human] complement is pronounced *who* (cf. Kayne 1994, fn. 12, pg. 154)

a. the [[$_{NP}$**student**]$_i$ [$_{CP}$ [which [$_{NP}$**student**]$_i$]$_j$ C^0 [Jonah likes t_j]]]
Pronounced: 'the student who Jonah likes'

b. the [[$_{NP}$**student**]$_i$ [$_{CP}$ [[which [$_{NP}$**student**]$_i$]'s brother]$_j$ C^0 [Jonah likes t_j]]]
Pronounced: 'the student whose brother Jonah likes'

c. the [[$_{NP}$**student**]$_i$ [$_{CP}$ [[[which [$_{NP}$**student**]$_i$]'s brother]'s roommate]$_j$ C^0 [Jonah likes t_j]]]
Pronounced: 'the student whose brother's roommate Jonah likes'

But even setting aside the cases involving possessor extraction, the movement postulated is unusual in that it is an instance of a projecting movement i.e. an instance of movement where the moving phrase and not the target projects. None of the instance of movement that we are familiar with, e.g. *wh*-movement, raising, passive movement, and head-raising, involve a projecting movement. In all these cases, it is the target that projects. Chomsky (1995, sect. 4.4.2), Chomsky (1998, sect. 5:pg. 51) build into the derivational mechanism that in cases of of movement, it is always the target that projects.

If the analysis that we are suggesting is correct, then projecting movements do exist. Finite relative clauses and the various kinds of reduced relatives involve projecting movements out of a variety of categories. It does not seem to be the case that assuming the projecting movements that I do leads to any empirical complications.

An additional question remains: what drives this movement? Let us consider an answer proposed by Iatridou, Anagnostopoulou, and Izvorski (1998), who also assume the existence of projecting movements. They argue that the movement of the relative XP in a finite relative targeting the relative C' is an instance of a projecting movement. The resulting object is nominal which they propose is why it is then able to combine with an NP. The movement of the Relative XP is triggered in order to make the relative clause nominal. It is plausible that the movement of the head NP out of the relative CP is for a similar reason. I will set aside the many other questions raised by the existence of projecting movements in general and the particular projecting movement that I need for further research.

The conceptual problems notwithstanding, this proposal has an advantage in that we do not need to stipulate anything special about the (non-)pronunciation of CP-internal material. As is usual in chains created by movement, the highest element in the chain is pronounced and the other elements deleted. We can therefore provide a straightforward explanation for the ungrammaticality of '*the *student*$_i$ [which *student*$_i$ Jonah likes]'.

If we assume that the relationship between the NP-head and the complement of the relative operator inside the CP/PrtP is mediated by movement, then there are two chains involved in the derivation of a relative clause.[20]

(60) a. the [[$_{NP}$ **book**]$_i$ [$_{CP}$ [which [$_{NP}$**book**]$_i$]$_j$ C^0 [Jonah likes [which book]$_j$]]]
Chains: Head NP chain, Relative phrase A'-chain

b. the [[$_{NP}$ **student**]$_i$ [$_{PrtP}$ [$_{NP}$**student**]$_i$ reading the new Rushdie book]]
Chains: Head NP chain, Trivial PrtP-internal chain

c. the [[$_{NP}$ **student**]$_i$ [$_{PrtP}$ [$_{NP}$**student**]$_i$ likely [[$_{NP}$*student*]$_i$ to win the race]]]
Chains: Head NP chain, PrtP-internal chain *A*-chain

The *A*′-chain in (60a) is interpreted via copy-deletion and Safir (1999) notes that the relative clause-internal *A*′-chain patterns with *A*′-chains in questions with respect to reconstruction. I assume that the PrtP-internal *A*-chain patterns with *A*-chains elsewhere with respect to reconstruction.[21] We now come to the behavior of the head NP chain with respect to interpretation/reconstruction. It has been noted by Safir (1999) that the material in the head NP does not pattern with other material in the [Spec,CP] of the relative clause with respect to reconstruction (cf. 61a vs. b). The non-CP-external material patterns with questions (cf. 61b, c).

(61) (ex. a from Sauerland 1998, b and c from Safir 1999)

a. I respect [[any depiction of John$_i$] [he$_i$'ll object to]].

b. *I respect [[any writer] [[whose depiction of John$_i$] he$_i$'ll object to]].

c. *Which pictures of Marsden$_i$ does he$_i$ display prominently?

The grammaticality of (61a) can be taken to show that the NP head *any depiction of John* can be interpreted outside the CP. If the head NP was always interpreted inside the CP, we would expect (61a) to be a Condition C violation like (61b, c). In (61a), we can interpret the CP-external copy of the head NP chain due to which the head NP does not have to be reconstructed following the conditions on *A*′-reconstruction. The option of CP-external interpretation is not available for *depiction of John* in (61b). The phrase *depiction of John* is only part of the *A*′-chain and not of the head NP chain. The conditions on *A*′-reconstruction force reconstruction of *depiction of John* and cause a Condn. C violation (cf. also 61c). Therefore we can conclude that the syntactic material of the head NP is at least sometimes interpreted CP-externally.

A′-reconstruction cases like (61) aside, it seems to be the case that interpreting the syntactic material of the head NP CP-externally and interpreting it in the highest link of the CP-internal *A*′-chain yields the same results. Interpreting the syntactic material of the head NP in lower copies of the *A*′-chain yields the effects of a 'head-raising' analysis and interpreting the syntactic material of the head NP in the highest link of the A'-chain yields (practically) the effects of the 'head-external' analysis. For simplicity, unless necessary, when computing the interpretation of relative clauses, I will ignore the option of CP-external interpretation of the head NP.

6.4. Reduced relatives and case

Our original analysis of reduced relatives did not involve lexical material in the subject position of the relative clause. The subject position was occupied by PRO, which does not need case (or it needs and receives null case). The analysis did not involve any nominals other than PRO in non-case-marked positions. So we did not have any problem with case-licensing. Let us consider the modified proposal from the perspective of case-licensing.

(62) the $[[_{NP}$**philosopher**$]_i$ $[[_{NP}$**philosopher**$]_i$ reading the Meinong text]]

The modified proposal involves the NP *philosopher* in the subject position of the reduced relative. Case is not licensed in the subject position of the reduced relative so the question of how the PrtP-internal NP *philosopher* is case-licensed arises. I will suggest two answers.

The first option assumes that it is only DPs that need to be case-licensed and not NPs.[22] This assumption does not seem to lead to any empirical problems since case theory seems to only be concerned with the distribution of nominal arguments which under plausible assumptions are always DPs. The phrase that is not case-licensed is an NP, which under these assumptions is not a problem. We proposed in footnote 20 that in finite relative clauses, there is always a relative operator - overt or covert. Since the relative operator is syntactically a determiner (i.e. it occupies D^0), the phrase that undergoes A'-movement is a DP and therefore the movement can only take place from case-marked positions. Thus we can explain the ungrammaticality of **the book Miguel is fond*. Another fact that falls out from assuming that NPs do not require case is the ungrammaticality of **the the philosopher reading the Meinong text*. The DP *the philosopher* would not be case-licensed inside the PrtP causing ungrammaticality.[23]

The second option assumes that NPs do need to be case-licensed. The NP in the subject position of the PrtP is case-licensed by licensing of case to a member of its chain: the PrtP-external head NP.

(63) The $[[_{NP}$**philosopher**$]_i$ $[[_{NP}$**philosopher**$]_i$ reading the Meinong text]] is obtuse.

The head NP which is external to the PrtP is the complement of *the*. It receives case from *the* and since it is in a chain with the subject of the PrtP, it case-licenses the subject of the PrtP. This kind of case-licensing is not able to rescue **the book Miguel is fond* because as per our assumption in footnote 20, we have two distinct chains, one involving *book* and one involving *Op book*. The CP-external NP head will case-license the CP-internal *book* but will not case-license *Op book*.

The two proposals seem to make similar predictions so far. They differ in that the relative clause-internal (possibly trivial) chain lacks case under the first proposal but has case under the second proposal. In the earlier version of our proposal, we connected the fact that only matrix subjects could be relativized in reduced relatives to the assumption that A'-chains need case and that since PRO does not receive case, it cannot be A'-moved. Since under the first option, the relative clause-internal chain in a reduced relative does not receive case, the locality facts follow straightforwardly. The existence of the locality facts under the second option is puzzling since the relative clause-internal chain does receive case via case transmission. Therefore, I will only consider the first option from this point onwards.

6.5. Interpreting the new proposal

How is the new proposal to be interpreted? An impressionistic treatment is provided here using a descriptive convention from Sauerland (1998) (see Bhatt (2002) for a more detailed treatment). As noted earlier, the CP-external head will be ignored for purposes of interpretation. The interpretation obtained by interpreting the CP/PrtP external head NP is the same as the interpretation obtained by interpreting the highest CP/PrtP-internal copy.

(64) Finite Relative Clauses

a. the [picture of Hafdis$_i$]$_j$ $[_{CP}$ [picture of Hafdis$_i$]$_j$ C^0 [she$_i$ likes [picture of Hafdis$_i$]$_j$]]
'Head-external analysis': interpret higher copy, no condition C:
LF: the [[λx [picture of Hafdis, x]] [λx [she likes x]]] (combine by Predicate Modification)
$\iota(\lambda x[pic(x,h) \wedge likes(h,x)])$

'Head-raising analysis': interpret lower copy, condition C violation
LF: the λx [she likes [picture of Hafdis, x]]

b. the [picture of himself$_i$]$_j$ [$_{CP}$ [picture of himself$_i$]$_j$ C^0 [Jonah$_i$ likes [picture of himself$_i$]$_j$]]
'Head-external analysis': interpret higher copy, leads to unbound *himself*
LF: the [[λx [picture of himself,x]] [λx [Jonah$_i$ likes x]]]
'Head-raising analysis': interpret lower copy, *himself* is properly bound
LF: the λx [Jonah$_i$ likes [picture of himself$_i$,x]]
with appropriate rule of interpretation (in Chapter xx)
$\iota(\lambda x[pic(x,j) \wedge likes(j,x)])$

(65) Reduced Relative
the [philosopher$_i$ [$_{PrtP}$ philosopher$_i$ reading Erewhon]]
LF: the [[λx [philosopher,x]] [reading Erewhon]]
(combine the two predicates by Predicate Modification)
$\iota(\lambda x[phil(x) \wedge reading(x,e)])$

The interpretation of the highest copy of the A'/A-chain inside the CP/PrtP respectively is formally identical to the head external analysis and does not seem to be any new problems. The interpretation of lower copies of the A'/A-chain inside the CP/PrtP respectively raises non-trivial problems that concern how the lower copy of the trace combines with its sister node, what the variable over which the relativization is taking place is, and what the resultant relative clause denotes and how it can be used. Some of these issue will be addressed in Chapter 3.

7. Conclusions

The major conclusion that we have arrived at in this chapter is that infinitival relative clauses form a heterogeneous class. In this they are unlike finite relative clauses, which are all CPs involving A'-movement of a relative operator. Subject infinitival relatives are reduced relatives; their formation does not involve A'-movement . They can be modal or non-modal. If they are modal, the source of the modality lies within the infinitival clause. Non-subject infinitival relatives are full relatives. Like finite relative clauses, they are always CPs.

Their formation involves A'-movement. They contrast with subject infinitival relatives in always being modal. The source of the modality is identified as the infinitival [+rel/wh] C^0.

We also argued that the head NP of a relative clause may originate inside the relative clause CP in both reduced and full relative clauses. This allows for reconstruction of the head NP into the relative clause in both full and reduced relative clauses. The derivation of full relative clauses involves CP-internal A'-movement and an operation of head externalization. The derivation of reduced relatives only involves the latter operation. The objects moved by A'-movement are DPs or larger while the objects targeted by the operation of head-externalization are NPs. We assume that DP-chains need case but not NP-chains. The absence of case in reduced relatives connects with the absence of A'-movement in reduced relatives and the fact that the abstraction in a reduced relative is always strictly local.

Chapter 3
Non-Modal subject infinitival relatives

In Chapter 2, subject infinitival relatives were analyzed as reduced relatives. It was noted that certain kinds of modal readings were generally available with subject infinitival relatives. In addition, a non-modal interpretation was also noted to be available in certain environments. In this chapter, we identify the environments where non-modal readings of infinitival relatives are available. It is proposed that in order for the infinitival clause to be interpreted non-modally, it has to be in a particular structural configuration with limited class of licensing elements. The analysis involves a 'head-raising' analysis of relative clauses. Using the semantics of the temporal superlative as probe, the interpretive properties of the infinitival clause that appears in non-modal infinitival relatives are explored in some detail.

1. Properties of non-modal infinitival relatives

Non-modal readings are only a possibility with subject infinitival relatives (cf. Kjellmer 1975, Geisler 1995, Pesetsky and Torrego 2001).

(66) a. This is [the best book ___ to appear until now (on this topic)]. (non-modal, subject gap)

b. This is [the best book to read ___]. (modal, object gap)

The fact that non-modal readings are unavailable with non-subject infinitival relatives (cf. 66a vs. b) follows from our analysis of object infinitival relative clauses as involving A'-movement to the [Spec,CP] of a C^0[+rel/wh,+inf] and the proposal that C^0[+rel/wh,+inf] has a modal interpretation (see chap. 2, sect. 3.1). Therefore if we have a non-subject infinitival relative, we have modality. Subject infinitival relatives are reduced relatives - there is no modal complementizer to automatically make them modal. The infinitival clause can be interpreted as modal if a covert modal is present. Nothing, however, forces a covert modal to be obligatorily present and therefore a non-modal reading is available in (66a).

1.1. Licensing the non-modal reading

The non-modal interpretation of the infinitival clause seems to be dependent upon the presence of one of the following kinds of modifiers: superlatives, ordinals, and *only* (cf. the contrast between 67 and 68). The examples in (67) are either ungrammatical altogether or are only grammatical under a modal reading.

(67) (# on the non-modal reading)

a. # (I took a picture of) every boy to climb Mt. Everest.

b. # (I took a picture of) most boys to climb Mt. Everest.

c. # (I took a picture of) some boy to climb Mt. Everest.

d. # (I took a picture of) a boy to climb Mt. Everest.

e. # (I took a picture of) no boy to climb Mt. Everest.

f. # (I took a picture of) the boy to climb Mt. Everest.

g. # (I took a picture of) the few boys to climb Mt. Everest.

h. # (I took a picture of) the many boys to climb Mt. Everest.

Superlatives, Ordinals and nominal *only* allow for a non-modal interpretation of the infinitival clause.

(68) a. (I took a picture of) the first person to climb Mt. Everest last year.

b. (I took a picture of) the only person to climb Mt. Everest last year.

c. (I took a picture of) the tallest person to climb Mt. Everest last year.

Ordinals can license non-modal readings from inside the Infinitival Relative as well as outside the Infinitival Relative.[24]

(69) a. the boy to climb Mt. Everest first/last

b. the first/last boy to climb Mt. Everest

In addition, cardinality adjectives like *two, few, many* in collocation with *one of the*) permit non-modal readings. [25]

(70) a. *few*
On Sept. 26, 1984, Mr. Gravitt sued, becoming [*one of* [the few citizens to use the statute in many years]].

Still, Mr. Feinberg is [*one of* [the few lawyers to make a rich and steady diet out of mediation, which accounts for 75 % of his practice]].

But its market share shrank – significantly in the case of mortgage-backed debt, [*one of* [the few areas to increase in volume last year]].

b. *many*

When Revlon Chairman Ronald Perelman recently hinted that he might try to acquire an industrial or manufacturing company, [*one of* [the many stocks to wiggle upward]] was Allied-Signal.

c. numeral

Alan Greenberg, chairman of Bear Stearns, [*one of* [the five dealers to curtail its program trading]], said that if the market's course has been smoother in the past two weeks, it was only "a coincidence," and not a result of the pullback.

The corresponding examples with *few/many*/numeral without overt marking of partitivity are degraded (compare 71a-c with 71d-f).

(71) a. ??? I have met [the many people to climb Mt. Everest last year].

b. ??? I have met [the few people to climb Mt. Everest last year].

c. ?? I have met [the two people to climb Mt. Everest last year].

d. I have met [the first person to climb Mt. Everest last year].

e. I have met [the only person to climb Mt. Everest last year].

f. I have met [the tallest person to climb Mt. Everest last year].

At present, I have no explanation for why there is a partitivity requirement for a non-modal reading of a subject infinitival relative with numerals, *few*, and *many*.

In addition, negation also seems to allow non-modal readings of the infinitival clause.

(72) (exs. 21, 22, pg. 326, Kjellmer (1975))

a. The captain is not the man to bother about such things.
($\approx$ the captain is not the man who bothers about such things)

b. He is not a man to let his friends down.
($\approx$ he is not a man who lets his friends down)

However, these cases have properties that are different from the cases seen in (68) and (70). The infinitival clause in the examples in (68) and (70) can receive a telic episodic interpretation. In contrast, non-modal readings licensed by a matrix negation are not compatible with a telic episodic interpretation of the infinitival clause (cf. 73). Only generic readings seem to be permitted in the examples in (72).

(73) a. # Neil Armstrong is not a/the man to walk on Mt. Everest.

b. # Microsoft's stock was not the stock to wiggle upward.

In addition, native speakers judge examples like (72) as marked/archaic. For this reason, in the discussion of non-modal infinitival relatives, we will set aside instances of licensing by a matrix negation as in (72).

The non-modal reading behaves like a polarity item as noted in Giannakidou (2003b)'s review. As noted above, its class of licensors includes superlatives, ordinals, and adnominal *only*. Syntactically represented partitivity together with *many*, *few*, and numerals in their modifier guise also licenses the non-modal interpretation. This class of licensors overlaps with licensors of more familiar polarity items like *ever* but is not co-extensive with it. That superlatives, ordinals, and adnominal *only* license *ever* is well-documented cf. Linebarger (1980). The role of partitivity in NP-licensing and even whether it has a role is unclear. On the other side, not everything that licenses *ever* licenses the modal reading. Downward entailing determiners such as *every* and *no* license *ever* but they do not license the non-modal interpretation.

Setting aside the role of partitivity, the following generalizations can be made at this stage. The first is that the elements that license the non-modal interpretation are never determiners - syntactically they behave like adjectival modifiers. The second is that these licensors seem to always co-occur with the definite determiner; this is a point that also holds of the partitive licensing cases. The first generalization is one that will follow from our proposal concerning the configuration in which the non-modal interpretation is licensed. The second will unfortunately not receive an explanation but will be shown to be a more general property of subject infinitival relatives in sect. 1.3. .

1.2. Non-Modal infinitival relatives and modal readings

Both modal and non-modal readings are easily available when the gap is in the subject position of a passive infinitive.

(74) This is the best argument to be considered by the committee.
$\approx$ (... which has been considered ...) (non-modal)
$\approx$ (... which is to be considered ...) (modal)

If the gap is in the subject position of an active infinitive, modal readings are harder to obtain, but not altogether absent (cf. 75).

(75) Joseph was the first/only/tallest man to walk on the moon last year.
$\approx$ Joseph was the first/only/tallest man who walked on the moon last year. (non-modal)
$\approx$ Joseph was the first/only/tallest man who was supposed to walk on the moon last year.

The availability of the modal reading is not surprising. We have seen in Chap. 2 that subject infinitival relatives can be modal, that the modality in subject infinitival relatives is part of the infinitival complement and that it comes in two flavors. (76) shows that the modality in a subject infinitival relative clause can be purpose-oriented or simply future-oriented.

(76) The man [to fix the sink] is here.
$\approx$ The man whose purpose it is to fix the sink is here.
$\approx$ The man who is scheduled/supposed to fix the sink is here.

If the infinitival clause is passive the future/schedule-oriented reading seems to be the only one available.

(77) The patient [to be examined for tomorrow's operation] is sleeping on the table.
$\approx$ The patient who is scheduled/supposed to be examined for tomorrow's operation is sleeping on the table. (future/schedule-oriented)
??$\approx$?? The patient whose purpose it is to be examined for tomorrow's operation is sleeping on the table. (purpose)

We propose that in general the infinitival clause that appears in subject infinitival relatives is ambiguous between an infinitival clause that is interpreted modally and an infinitival clause that is interpreted non-modally. The infinitival clause that is interpreted modally does not have special licensing conditions and can appear wherever subject infinitival relatives appear. Hence it can appear both in (74) and (76). The infinitival clause that is interpreted non-modally, however, needs to be licensed and can therefore appear only in certain environments. While it can appear in (74), it cannot appear in (76).

The net result is that infinitival relative clauses that permit non-modal interpretations also allow modal interpretations.

1.3. Restrictions on modal subject infinitival relatives

We have seen that subject infinitival relative clauses which allow for a non-modal interpretation also allow for modal interpretations. One component of the licensing of the non-modal interpretation is that there is a definite determiner along with the adjectival licensor of the non-modal interpretation. There are also subject infinitival relatives that permit only modal interpretations - see (76, 77). Somewhat unexpectedly, it seems that even such subject infinitival relatives need a definite determiner for wellformedness. Giannakidou (2003b) reports that her informants find the following ungrammatical.

(78) a. *I took a picture of every boy/most boys/three boys/ few boys/ many boys/a boy [to climb Mt. Everest].

b. *I just saw a man/every man/most men/few men/some man/ three men [to fix the sink].

It seems then that it is a general property of subject infinitival relatives that they require the definite determiner for wellformedness and not just a property of non-modal infinitival relatives. Unfortunately we do not know why this should be so.

In the following sections, we will examine the properties of elements license the non-modal interpretation. The question of the configuration required for the licensing of the non-modal interpretation will also be addressed. But before we do that it is worth articulating a background assumption. We have been talking about licensing a non-modal interpretation. But what exactly is licensed when the non-modal interpretation is licensed? We will assume that there is a syntactic element, a head for concreteness, which corresponds to the non-modal interpretation and which is what needs to be licensed to get the non-modal interpretation. We further assume that this head is in complementary distribution with a head which contributes modal meaning in infinitives. When no licensor is present, the syntactic position still needs to be realized and so the head contributing a modal interpretation appears.

2. A raising analysis of non-modal infinitival relatives

We will show that the licensing configuration for Non-modal Infinitival Relatives is obtained via 'reconstruction' of the head NP into the subject position of the infinitival clause and subsequent movement of an element out of the 'reconstructed' head NP to a position from where it licenses the non-modal interpretation of the infinitival clause. Following the discussion in Chapter 2, subject infinitival relatives are reduced relatives and we will use the analysis of reduced relatives developed there. Recall that this analysis allows for reconstruction into the 'gap' inside the relative clause.[26]

2.1. Raising analysis of relative clauses

The main idea within the raising analysis of relative clauses is that the 'head NP' starts off within the relative clause and is 'promoted' to the matrix clause via movement. A summary of the arguments for a raising analysis of relative clauses is provided in chap. 2, sect. 5. Here, we quickly repeat some of the arguments that have been made to motivate the raising analysis of relative clauses (cf. Kuroda 1968, Schachter 1973, Vergnaud 1974, Brame 1976, Carlson 1977, Kayne 1994, Sauerland 1998, and Bhatt 2002; for a different treatment see Sharvit (1999)).

The first argument comes from binding. If for an R-expression to bind an anaphor or for a quantifier to bind a pronoun, the former should c-command the latter at LF, then examples like the following pose a problem for the head-external analysis of relative clauses. In (79a-c), we have instances of an element inside the relative clause binding an anaphor/pronoun in the head of the relative clause. Since relative clauses are islands for overt movement and QR moving the subject of the relative clause out of the relative clause is not an attractive option.

(79) a. Mary bought the picture of himself$_j$ that Bill$_j$ saw. (Kayne 1994)

b. The interest in [each other]$_i$ that [John and Mary]$_i$ showed was fleeting.
(Schachter 1973, 43a)

c. The book on her$_i$ desk that every professor$_i$ liked best concerned Model Theory.
(Sauerland 1998, 63:52c)

Under the raising analysis of relative clauses, there is a copy of the head NP inside the relative clause that the binder in (79) c-commands. If we interpret this copy at LF, we have an appropriate configuration at which the binding conditions are met.

The second argument comes from the domain of the interpretation of bound phrasal expressions and complex phrasal idioms. This argument is based on the assumption that certain lexical items have to be in a local enough configuration at some level of the derivation for them to receive their idiomatic interpretation. The level of derivation that matters can be the point of merger (D-structure in earlier terminology) or LF. In order for the idiomatic interpretation to be available, the various parts of a complex phrasal idiom must be merged sufficiently close to each other. In addition, there may also be locality considerations at LF. (80) has been used to argue that the photography interpretation of *take a picture* is only available if *take a picture* forms a unit at LF.[27]

(80) (Ex. a from Chomsky 1993)

a. Which picture of $\text{himself}_{i/j}$ does John_i think that Bill_j took?
Claim: Photography reading is unavailable if *himself* refers to *John*.

b. Which picture of himself_i does John_i think that Mary took?
Claim: Photography reading is unavailable

Given the above assumptions about the locality requirements on the interpretation of complex phrasal idioms, the availability of an idiomatic interpretation in (81) is puzzling on the head external analysis. Part of the phrasal idiom is outside the relative clause in (81a, b) and there is no relationship, local or non-local, between *make* and *headway* and *take* and *pictures* respectively.

(81) from Brame (1976)

a. The headway John made proved insufficient.

b. All the pictures John took showed the baby.

The right configuration is available under the Raising analysis. The head NP starts off as an argument of *make/take* satisfying the locality requirement. If in addition the locality requirement holds at LF, we can satisfy this locality requirement by interpreting an appropriately local copy.

2.2. Interpreting the raising analysis (for reduced relatives)

For the purposes of subject infinitival relatives and reduced relatives in general, I will make the simplifying assumption that only the copy of the head in the subject position is interpreted. The *in-situ* head *book* in (82a) combines with the predicate *reading the Meinong text* via Predicate Modification as shown in (82b).

(82) a. the [[$_{NP}$ student]$_i$ [$_{PrtP}$ [$_{NP}$ student]$_i$ reading the Meinong text]]
b. the [[$\lambda x[student(x)]$] [$\lambda x[reading(x, mt)]$]]
Predicate Modification: the [$\lambda x[student(x) \wedge reading(x, mt)]$]

Since we will not be interpreting anything but the highest PrtP-internal copy in our discussion of Non-modal Infinitival Relatives, the above treatment of the semantics of raising relative clauses will suffice. Many non-trivial questions arise if we interpret any copy other than the highest one but we will set them aside here. See Bhatt (2002) for details.

2.3. Raising analysis applied to non-modal infinitival relatives

By interpreting the highest IP-internal copy of the head NP and moving the degree head *est* from out of the NP and adjoining it to the IP, we get the LF in (83).

(83) The tallest person to walk on the moon

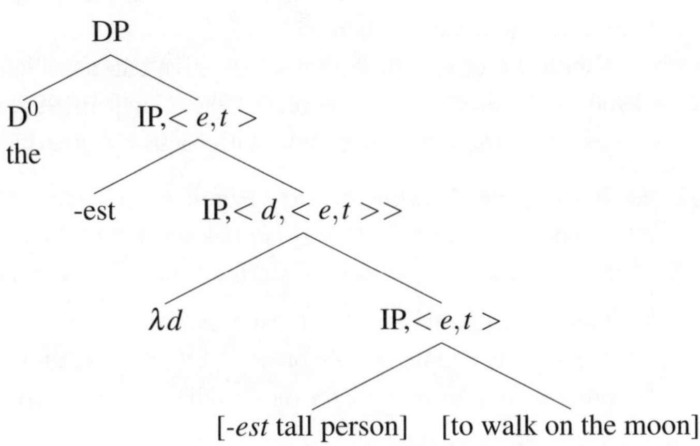

LF proposed: the [*est* λd [[*d*-tall person] to walk on the moon]]

For interpretive purposes, the IP-external head NP will be ignored. Since we have a Raising analysis, the head NP *tallest person* can be interpreted inside the relative clause and it is its (highest) relative clause internal copy that we will use for interpretation.

At LF, *-est* moves out of the relative clause internal copy of the head NP and takes scope over the entire clause. In doing so, it is not alone: *first* and *only* also display similar behavior. We will see some arguments for this movement in sect. 3.

The arguments for adopting (83) as the proper analysis of non-modal infinitival relative clauses will be presented in the sections that follow. To preview, the above structure will provide us with a structural characterization of the class of modifiers that permit non-modal interpretation of infinitival relatives. (Details in sect. 4.1). It will also help us to provide an analysis of facts related to the temporal interpretation of the associated relative clause (Details in sect. 6).

3. Motivating movement of *est*/ordinals/*only*

The analysis of non-modal infinitival relatives proposed in the previous section assumed a raising analysis of relatives clauses and movement of a degree operator/*only* out of the subject NP to an IP-adjoined position. The raising analysis of relative clauses has been argued for in the previous chapter and in this chapter. The proposal that a degree operator/*only* can and does move in the manner indicated needs to justified. This section will go over the set of facts that argue for such a movement.[28]

We begin with the observation that superlative can associate with focus (cf. Ross 1964, Jackendoff 1972, Szabolcsi 1986, Gawron 1995, Heim 1995). (84) provides an example of association with focus by superlatives.

(84) a. **Joan**$_F$ gave Caterina the most expensive present.
≈ Some people gave Caterina presents. Of all those presents, the present that was given by Caterina was the most expensive.

b. Joan gave **Caterina** $_F$ the most expensive present.
≈ Joan gave some people present. Of all those presents, the present Joan gave to Caterina was the most expensive.

Depending upon where the focus falls, we get truth-conditionally distinct readings (84a vs. b).

Szabolsci (1986) noted that the existence of locality constraints on the relationship between the superlative and the focussed element. The set of examples that motivated her observation are shown in (85). These examples show that superlatives can only associate with Focus/Wh when the Focus/Wh(-trace) is within a certain local domain containing the superlative. The superlative and the focussed element with which it associates can be in separate clauses (cf. 85a-d) and no c-command requirement holds between the superlative and the focussed element with which it associates (cf. 85a-c, vs. d). However, the superlative and the focussed element with which it associates cannot be separated by a finite clause (cf. 85e-g). For some speakers, the reading indicated to be absent in (85g), which involves association across a finite clause but not an island, is marginally available. The judgements for (85e-f), which involve association across an island, hold for all speakers.

(85) (* indicates the unavailability of the indicated reading)

a. **Bill**$_F$ expected [PRO to get the fewest letters].
$\approx$ Bill expected to get 2 letters, Jane expected to get 3, Polly expected to get 5 letters. Hence Bill expected to get the fewest letters.

b. **Bill**$_F$ wanted [PRO to climb the highest mountain].
$\approx_{dedicto}$ Bill wanted to climb a mtn. of height 10kms., Jane wanted to climb a mtn. of height 7kms., Polly wanted to climb a mtn. of height 5kms. Hence Bill expected to climb the highest mountain.

c. **Bill**$_F$ expected [Mary to get the fewest letters].
$\approx$ Bill expected Mary to get 2 letters, Jane expected Mary to get 3, Polly expected Mary to get 5 letters. Hence Bill expected Mary to get the fewest letters.

d. You expected the fewest people to like **Bill**$_F$.
$\approx$ You expected 2 people to like Bill, 3 people to like Jane, 5 people to like Polly. Hence you expected the fewest people to like Bill.

e. ***Bill**$_F$ likes the student who John gave the fewest presents.
$\neq$ Bill likes the student who John gave 10 presents. Jane likes the student who John gave 15 presents. Polly likes the student who John gave 20 presents. John also gave Einar 3 presents, but nobody seems to like Einar. (There may also be students who

John gave no presents). Hence Bill likes the student who John gave the fewest presents.

f. ***Bill**$_F$ made the claim that you got the fewest letters.
$\neq$ Bill made the claim that you got 2 letters, Jane made the claim that you got 3, Polly made the claim that got 5. Hence Bill made the claim that you got the fewest letters.

g. ***Bill**$_F$ said that you got the fewest letters.
$\neq$ Bill said that you got 2 letters, Jane said that yot 3, Polly said that you got 5. Hence Bill said that you got the fewest letters.

Szabolsci (1986) (with a few modifications) provides the following analysis for these facts: The background assumption is that superlatives are complex objects. e.g. the superlative *tallest* is composed of the adjective *tall* and the superlative morpheme *est*. Szabolcsi proposes that in order to associate with Focus/Wh, *est* must take scope over the focussed element/ *wh*-trace. This scope requirement is not met at S-structure (cf. 85d) but at LF. *est* moves at LF to a position where it adjoins to a clausal projection, the exact nature/semantic type of which depends upon the particular implementation. The movement of *est* is very constrained - not only can it not move out of islands (cf. 85e, f), it also cannot move out of a finite clause (cf. 85g).

The LF representations corresponding to (85b) and (85e) are shown in (86a, and b) respectively.

(86) a. ***Bill**$_F$ likes the student who got the fewest letters.
LF1: (assumes *est* takes a clausal argument, does not assume focus movement)
est λd [**Bill**$_F$ likes the student who got d-few letters]
bad due to island violating movement of *est*
LF2: (assumes *est* targets a predicate created by *wh*-movement or focus movement)
Bill$_F$ *est* λd [λx [x likes the student who got d-few letters]]
bad due to island violating movement of *est*

b. **Bill**$_F$ wanted to climb the highest mountain.
LF1: *est*λd [**Bill**$_F$ wanted to climb a d-high mountain]
LF2: **Bill**$_F$ *est* λd [λx [x wanted to climb a d-high mountain]]

In (86b), the degree operator *est* takes scope over *want* while the NP associated with it takes scope under *want*. Such instances of 'split-scope' argue for an approach like the current one where the *est* can move independently of the NP it is associated with.

3.1. A semantics for superlatives

We present here two proposals for the semantics of superlatives. In both proposals, *est* takes a contextual restriction as an argument. The value of this contextual restriction is determined by the contribution of focus. The proposals differ in whether *est* takes an external argument or not.

Under the two argument analysis, *est* takes as arguments an individual (the external argument) and a predicate of individuals and degrees.

(87) $est(C)(R)(x) \Longleftrightarrow \exists d.[x \in C \wedge R(x,d) \wedge \forall y[y \neq x \wedge y \in C \Longrightarrow \neg R(y,d)]]$
C is a contextual restriction, R is a predicate of individuals and degrees, x is an individual. (cf. Heim 1995, also see Szabolsci 1986)
$est(C)(R)(x)$ iff x is R to a degree that nothing else in C is.

A semantics based on Seuren (1978) is assumed for gradable adjectives. According to this semantics the following entailment holds:
x's height is $H \rightarrow \forall d[0 < d \leq H \rightarrow tall(d,x)]$
i.e. x is tall to every degree that is less than or equal to x's height. It follows from the above that if x is tall to a degree d, x is tall to every degree d' less than d.

I will assume that when superlative associate with focus, *est* takes the focussed element as its external argument. When the association is with *wh*-movement, the predicate created by movement of the *wh*-phrase is the internal argument of *est* and the *wh*-phrase the external argument. C contains the alternatives created by Focus/ *wh*-movement. So in (88a), C contains the contextually salient alternatives to *John*, and in (88b) to *Mary*.

(88) a. **John**$_F$ wrote the longest letter to Mary.
John [C-est] λd [wrote A d-long letter to Mary]
$est(C)(\lambda d \lambda x \exists l.[wrote(x,l,m) \wedge long(d,l)])(j)$
John has the property of writing a d-long letter to a degree nothing else in C does.

b. John wrote the longest letter to **Mary**$_F$.
QR the focussed constituent '**Mary**'

Mary λx [John wrote [C-est] long letter to x]
Move *est*
Mary [C-est] $\lambda d\lambda x$ [John wrote [C-est] long letter to x]
$est(C)(\lambda d\lambda x\exists l.[wrote(j,l,x)\wedge long(l,d)])(m)$
Mary has the property of having a *d*-long letter written to her by John to a degree that nothing else in *C* does.

The two argument analysis relies in (88b) on the movement of the focussed constituent to create its two arguments. We need to ensure that in (88b) the *est* takes as its external argument the focussed constituent, and not the unfocussed subject *John*. If this is not done, we lose the correlation between the placement of focus and the observed truth-conditions.

The one argument analysis takes as an argument a proposition. This proposition contains the focussed element which we would have needed to move in order to generate the arguments for the two argument analysis. *est* combines with a property of degrees and contextual restricted set of properties of degrees C, which is constrained by focus cf. Heim (1995).

(89) $est(C)(P)\leftrightarrow\exists d[P(d)\wedge\forall Q[Q\neq P\wedge Q\in C\Longrightarrow\neg Q(d)]]$
'There is a degree in *P* that is not in any other property of degrees in *C*.'
A semantics for gradable adjectives based on Seuren (1978) is assumed (cf. 87).

(90) **John**$_F$ is the tallest.
$P=\lambda d.tall(j,d)$
$C=\{\lambda d.tall(j,d),\lambda d.tall(k,d),\lambda d.tall(l,d),\ldots\}$
'There is a degree in *P* (e.g. the degree *h* corresponding to John's height) s.t. it is not in the other members of *C* i.e. *K* is not *h*-tall, *L* is not *h*-tall, etc.'

On the side of the one-argument analysis of *est* is its non-reliance on the movement of focussed constituents. The contextual restriction is obtained by accessing the alternatives evoked by focus. However, the one-argument analysis runs into a problem with relative clauses. Szabolcsi (1986) points out that 'association with focus' style effects appear with superlatives in relative clauses even in the absence of any overt focus.

(91) a. We should console the girl who got the fewest letters.
b. I have seen the girl from whom Peter got the fewest letters.

In (91), the contextual alternatives for the interpretation of *est* come from the set *girl* which is not focussed. If we want to retain the one-argument analysis, we have to assume that movement is able to provide alternatives in the same way as focus. This could be done, for example, by assuming that *wh*-traces are focussed. However, there are problems with this assumption - movement seems to be unable to create alternatives for *only*. cf. the difference between 'John only gave **Mary**$_F$ a present' vs. 'the girl who John only gave a present'. A two argument analysis does not run into a problem with relative clauses because it does not exclusively depend upon focus to generate its clausal argument.

To sum up, there seem to be arguments both for a one-argument and for a two-argument analysis of *est*. It will turn that for the analysis proposed for non-modal infinitival relatives, a two-argument analysis is the only one feasible. In addition, the one-argument analysis for finite relative clauses requires making the use of focussed traces, a device that is empirically problematic. Therefore, I will assume the two-argument from now onwards.

3.2. Focus-sensitivity of ordinals and nominal *only*

Like *est* and like their adverbial counterparts, nominal ordinals and nominal *only* also show focus-sensitivity. The focus-sensitive behavior of adverbial *first* is shown in (92) and the focus-sensitive behavior of nominal *first* is shown in (93).[29]

(92) a. **Joan**$_F$ hugged Mary first.
≈ Joan hugged Mary before anyone else hugged Mary.
LF1: *first* [**Joan**$_F$ hugged Mary] (one-argument analysis)
LF2: **Joan**$_F$ *first* λx [*x* hugged Mary] (two-argument analysis)

b. Joan hugged **Mary**$_F$ first.
≈ Joan hugged Mary before hugging anyone else.
LF1: *first* [Joan hugged **Mary**$_F$] (one-argument analysis)
LF2: **Mary**$_F$ *first* λx [Joan hugged *x*] (two-argument analysis)

(93) a. **Joan**$_F$ gave Mary the first telescope.
≈ Mary got a telescope from Joan before she got a telescope from anyone else.

b. Joan gave **Mary**$_F$ the first telescope.

≈ Joan gave a telescope to Mary before she gave a telescope to anyone else.

The focus-sensitive behavior of adverbial *only* (cf. Horn 1969, Anderson 1972, Rooth 1985, 1992) is shown in (94) and the corresponding focus-sensitive behavior of nominal *only* is shown in (95).[30]

(94) a. John only introduced **Bill**$_F$ to Sue.
≈ Out of the relevant set of people, John introduced Bill to Sue and he did not introduce anyone else to Sue.
LF1: *only* [John only introduced **Bill**$_F$ to Sue] (one-argument analysis)
LF2: **Bill**$_F$ *only* λx [[John only introduced x to Sue] (two-argument analysis)

b. John only introduced Bill to **Sue**$_F$.
≈ Out of the relevant set of people, John introduced Bill to Sue and he did not introduce Bill to anyone else.
LF1: *only* [John only introduced Bill to **Sue**$_F$] (one-argument analysis)
LF2: **Sue**$_F$ *only* λx [[John only introduced Bill to x] (two-argument analysis)

Association with focus readings of nominal *only* are harder to get but are available, at least for some speakers.

(95) a. **Joan**$_F$ gave Mary the only telescope.
≈ Some people gave Mary presents. Out of these presents, the only telescope that Mary got was from Joan. Joan/Other people may have given people other than Mary telescopes.

b. Joan gave **Mary**$_F$ the only telescope.
≈ Joan gave some people presents. Out of these presents, the only telescope that Joan gave was to Mary. Mary/Other people may have received telescopes from people other than Joan.

If we assume that for *first/only* to associate with a focussed element, *first/only* must c-command the relevant focussed element at LF, the arguments for *est* movement carry over to ordinals and nominal *only* too.

It is clear from the LFs indicated that the adverbial ordinals/*only* semantically combine with a clausal/VP constituent. It seems to be the case that nominal ordinals/*only* also combine with a clausal constituent. Their interpretation seems to indicate that they have the same semantics as their adverbial counterparts. The LFs corresponding to (93) and (95) are indicated in (96).

(96) a. **Joan**$_F$ gave Mary the first telescope.
LF1: *first* [**Joan**$_F$ gave Mary A telescope]
LF2: **Joan**$_F$ *first* λx [x gave Mary A telescope]

b. Joan gave **Mary**$_F$ the first telescope.
LF1: *first* [Joan gave **Mary**$_F$ A telescope]
LF2: **Mary**$_F$ *first* λx [Joan gave x A telescope]

c. **Joan**$_F$ gave Mary the only telescope.
LF1: *only* [**Joan**$_F$ gave Mary A telescope]
LF2: **Joan**$_F$ *only* λx [x gave Mary A telescope]

d. Joan gave **Mary**$_F$ the only telescope.
LF1: *only* [Joan gave **Mary**$_F$ A telescope]
LF2: **Mary**$_F$ *only* λx [Joan gave x A telescope]

Like *est*, nominal ordinals/*only* (when they associate with focus) combine with a clausal/VP constituent. They differ from *est* in that it is not clear what, if any, variable they are binding inside the NP where they originate. *est* binds a degree variable in the NP where it originates. On the other hand, it is not clear what variable nominal *first/only* bind. A movement that does not leave behind a trace is conceptually problematic.

However, the cases in (96) are not the only case where we seem to need a movement that does not leave behind a trace. One case of such a movement which has been the subject of some discussion in the recent literature (cf. Bolinger 1967, Stump 1981, Larson 1998) is shown in (97).[31]

(97) An occasional sailor strolled by.
'external' adverbial reading: Occasionally, a sailor strolled by.
LF: Occasionally [A sailor strolled by]

Another case is noted in (98) (cf. den Dikken, Larson, & Ludlow 1997). (98) has a reading where an *only* that seems to be associated in overt syntax with the NP *one banana* takes scope at LF away from it.[32]

(98) Jonah needs only one banana.
LF: only [Jonah needs one banana]

Like in the cases that we have been discussing, we need to move an element without leaving behind a trace. For the purposes of the discussion that follows, I will not have nominal ordinals/*only* not leave behind any trace. I assume that further discussion of these issues will either reveal the existence of a variable bound by the moved element or explicate why certain movements that do not leave variables behind are legitimate.

4. More on the licensing of the non-modal interpretation

4.1. Structural characterization of superlatives, ordinals, and *only*

Under the current proposal, superlatives, ordinals, and *only* when they occur with Non-modal Infinitival Relatives appear in the configuration in (99). I propose that this is the configuration in which a non-modal reading of the infinitival clause is licensed.

(99) *first/only/-est* takes the infinitival clause as a complement:

a. the [first [[person] to solve the problem]]

b. the [only [[person] to solve the problem]]

c. the [-est λd[[d-tall person] to solve the problem]]

The common feature shared by superlatives, ordinals, and *only* which allows the non-modal reading of the infinitival clause to be licensed is the fact that they are able to appear in the configuration shown in (99). They occur with *the* showing that they occur low in the NP structure - which is why they are able to reconstruct in contrast to quantifiers like *every*, *some*, *most* which occur in D^0 and so do not reconstruct. Recall that it is only the NP and not the DP that originates inside the relative clause. As such, only the NP and not the DP can reconstruct. Due to their particular semantics and syntax, they can occur as a sister to the infinitival clause in the configuration shown in (99).

The relevance of the configuration in (99) is also suggested by the fact that not all modifiers that can occur with *the* yield non-modal readings with Infinitival Relatives. Occurring with *the* means that a modifier can be interpreted as part of the NP and hence inside the relative clause. However, for the licensing configuration to obtain, the relevant modifier has to move out of the subject NP and take the infinitival clause as a complement. This is what happens with ordinals, superlatives, and nominal *only* and does not happen with *many*, *few*, and numerals (e.g. *two*).

many, few, two can occur under *the*:

(100) a. The many people who came to Bill's class left early.
(≈ Many_{weak} people came to Bill's class. They left early.)

b. The few people who came to Bill's class left early.
(≈ Few_{weak} people came to Bill's class. They left early.)

c. The two people who came to Bill's class left early.
(≈ Two_{weak} people came to Bill's class. They left early.)

Presumably they can reconstruct (see Bhatt (2002) for evidence that they do reconstruct). However, non-modal interpretations of the infinitival clause are not acceptable with *many, few, two*:[33]

(101) a. # Yesterday, I met the many people to climb Mt. Everest last year.

b. # Yesterday, I met the few people to climb Mt. Everest last year.

c. # Yesterday, I met the two people to climb Mt. Everest last year.

many, *few*, and numerals like *two* when they occur with *the* occur low in the NP structure combine with the rest of the NP like adjectives (cf. 102).

(102) $\iota\lambda x.P(x)$ returns the maximal individual that satisfies P if there is such a maximal individual.

a. The two books Bill likes
$\iota(\lambda x.[two(x) \wedge books(x) \wedge like(b,x)])$

b. The few books Bill likes
$\iota(\lambda x.[few(x) \wedge books(x) \wedge like(b,x)])$

c. The many books Bill likes
$\iota(\lambda x.[many(x) \wedge books(x) \wedge like(b,x)])$

Unlike the focus-sensitive ordinals, superlatives, and *only*, *few/many* and numerals do not have the option of moving out of the NP and taking a clausal complement. Even if they could in principle move out, they would not take the infinitival clause as a complement: since numerals/*few, many* have the semantic type $< e,t >$, they would combine with the infinitival clause which is of type $< e,t >$ by Predicate Modification and not by taking the infinitival clause as a complement.[34] Consequently, numerals/*few, many* are not in the right configuration to license a non-modal interpretation of the Infinitival Relative.

4.2. Licensing from inside the infinitival clause

The proposal for the configuration which licenses the non-modal interpretation of the infinitival clause receives support from the fact that *first/last* can license the non-modal interpretation from inside the infinitival clause also (cf. 103).

(103) a. The student to finish the marathon first got the gold medal.
(≈ The first student to finish the marathon got the gold medal.)

b. The student to finish the marathon last got a consolation prize.
(≈ The last student to finish the marathon a consolation prize.)

first/last internal to an infinitival clause can license the Non-Modal IR because at LF they are in the same position as the 'lowered' nominal modifiers *first/last*, directionality aside.

(104) the *first* [[student] to finish the marathon]]

We have a simple explanation for the availability of a non-modal interpretation in (103). The question now arises why a non-modal reading is not possible in (105). There seems to be a licensor and yet a non-modal reading seems unavailable.

(105) a. # The student to only climb Mt. Everest

b. # The student to scream the loudest

c. # The boy to give Mary the only telescope

The unavailability of (105a) can be explained straightforwardly. Adverbial *only* only associates with material in its syntactic scope. It does not seem to move covertly. This can be seen in the contrast the association with focus possibilities between (106a) and (106b).

(106) a. **John**$_F$ only introduced Bill to Sally.
No association of *only* with focus

b. Only **John**$_F$ introduced Bill to Sally.
Association of *only* with focus

Since the movement site in (105a) is not in the syntactic scope of *only*, *only* cannot associate with it i.e. *only* cannot appear in the licensing configuration.

The deviance of (105b, c) still needs to be accounted for. The deviance of (105b, c) shows that the licensing configuration as it stands needs to be strengthened. I sill suggest two ways of adding an additional constraint to the licensing configuration so that it able to account for the deviance of (105b, c).[35]

The first way involves adding a surface structure constraint on the licensing of the non-modally interpreted infinitival clause. We know from the literature on NPI-licensing (cf. Ladusaw 1980 among many others) that the licensing of NPIs seems to involve both a surface structure c-command component and a LF component. It is plausible that the licensing of the non-modally interpreted infinitival clause similarly involves a surface structure c-command component. If so, the cases in (105b, c) would be ruled out due to lack of surface structure c-command between the licensor and the infinitival clause.

Recent work within generative grammar has attempted to state all constraints as either constraints on the derivational mechanism or as constraints on LF (or some other abstract level). A constraint like 'c-command at S-structure' is conceptually problematic from the perspective of these attempts. Therefore, a way of strengthening the licensing configuration that only involves LF-constraints is now suggested.

If we compare the LFs of (107a) and (107b), we see that *est* binds degree variables in different locations. In contrast, in (103) nominal and adverbial *first* yield essentially the same LF configuration.

(107) a. # the boy to scream the loudest
the *est* λd [[boy] to scream *d*-loud]

b. the tallest boy to scream
the *est* λd [[*d*-tall-boy] to scream]

We suspect that this difference is what is responsible for the oddness of (107a = 105b) and (105c). If this is the case the licensing configuration has to be strengthened to include reference to a relationship between the licensor and the subject of the infinitival clause i.e. the licensor should bind a variable in the subject position. The question of what kind of variable *first/only* bind in the NP where they originate comes up here. As before, we set this question aside for now.

5. Loss of focus-sensitivity with non-modal infinitival relatives

The class of modifiers that seem to be necessary for the Non-modal interpretation are all focus-sensitive i.e. they can associate with focus. Under the proposed analysis, the focus-sensitive element - *only*, ordinals, *est* must be interpreted inside the relative clause to be in the right configuration for the non-modal reading of the Infinitival Relative to obtain.

(108) (recall that the head NP is not being interpreted)

a. the [$_{IP}$ first [$_{IP}$ [person] to solve the problem]]

b. the [$_{IP}$ only [$_{IP}$ [person] to solve the problem]]

c. the [$_{IP}$ est λd [$_{IP}$ [d-tall person] to solve the problem]]

If the focus-sensitive element stays put inside the Relative Clause, it will be unable to associate with focussed phrases in the matrix clause since they will be outside its scope (cf. sect. 3). Further movement of the focus-sensitive element will be blocked by constraints on movement since Relative Clauses are islands. So the prediction is that an ordinal/superlative/*only* that is with a Non-Modal Infinitival Relative will not be able to associate with Focus. We will see in the following sections that this prediction is borne out.

5.1. Loss of focus-sensitivity of superlatives, *only*, ordinals

(109) provides an example of association with focus by superlatives (cf. Ross 1964, Jackendoff 1972, Szabolcsi 1986, Gawron 1995, Heim 1995). [36]

(109) a. **Joan**$_F$ gave Mary the most expensive telescope.
$\approx$ Some people gave Mary telescopes. Of all those telescopes, the telescope that was given by Joan was the most expensive.

b. Joan gave **Mary**$_F$ the most expensive telescope.
$\approx$ Joan gave some people telescopes. Of all those telescopes, the telescope Joan gave to Mary was the most expensive.

Depending upon where the focus falls, we get truth-conditionally distinct readings (109a vs. b). If, however, there is a non-modally interpreted infinitival relative associated with the superlative, the truth-conditionally distinct readings disappear.

(110) a. **Joan**$_F$ gave Mary the most expensive telescope [to be built in the 9th century].
$\approx$ It was Joan who gave Mary the most expensive telescope [to be built in the 9th century].

b. Joan gave **Mary**$_F$ the most expensive telescope [to be built in the 9th century].
≈ It was Mary who Joan gave the most expensive telescope [to be built in the 9th century].

The two readings (110a vs. b) differ in terms of presuppositions but seem truth-conditionally identical. The superlative *most expensive* does not associate with focus in the matrix clause (compare with 109). Paraphrases for the missing readings are provided below in (111).

(111) a. **Joan**$_F$ gave Mary the most expensive telescope [to be built in the 9th century].
≠ Some people gave Mary telescopes built in the 9th century. Of all those telescopes, the telescope that was given by Joan was the most expensive.

b. Joan gave **Mary**$_F$ the most expensive telescope [to be built in the 9th century].
≠ Joan gave some people telescopes built in the 9th century. Of all those telescopes, the telescope given to Mary was the most expensive.

The paraphrases are not pragmatically odd and the judgement is quite clear that these readings are clearly absent.

It has been noted earlier that adverbial ordinals (112) and nominal ordinals (113) associate with focus leading to truth-conditionally distinct readings.

(112) a. **Joan**$_F$ hugged Mary first.
≈ Joan hugged Mary before anyone else hugged Mary.

b. Joan hugged **Mary**$_F$ first.
≈ Joan hugged Mary before hugging anyone else.

Ordinals as nominal modifiers are also focus-sensitive

(113) a. **Joan**$_F$ gave Mary the first telescope.
≈ Mary got a telescope from Joan before she got a telescope from anyone else.

b. Joan gave **Mary**$_F$ the first telescope.
≈ Joan gave a telescope to Mary before she gave a telescope to anyone else.

The two readings in (114a, b) differ only presuppositionally and not truth-conditionally (compare with 113a, b).

(114) a. **Joan**$_F$ gave Mary the first telescope [to be built in the 9th century].
≈ It was Joan who gave Mary the first telescope [to be built in the 9th century].

b. Joan gave **Mary**$_F$ the first telescope [to be built in the 9th century].
≈ It was to Mary that Joan gave the first telescope [to be built in the 9th century].

Paraphrases for the missing readings are provided below in (115).

(115) a. **Joan**$_F$ gave Mary the first telescope [to be built in the 9th century].
≠ Mary got a telescope built in the 9th century from Joan before she got a telescope built in the 9th century from anyone else.

b. Joan gave **Mary**$_F$ the first telescope [to be built in the 9th century].
≠ Joan gave Mary a telescope built in the 9th century before she gave a telescope built in the 9th century to anybody else.

The focus-sensitivity of adverbial *only* has been the topic of much study (cf. Horn 1969, Anderson 1972, Rooth 1985, 1992).

(116) a. John only introduced **Bill**$_F$ to Sue.
≈ Out of the relevant set of people, John introduced Bill to Sue and he did not introduce anyone else to Sue.

b. John only introduced Bill to **Sue**$_F$.
≈ Out of the relevant set of people, John introduced Bill to Sue and he did not introduce Bill to anyone else.

Nominal *only* also displays association with focus effects though these seem to be tougher to get.

(117) a. **Joan**$_F$ gave Mary the only telescope.
≈ Some people gave Mary presents. Out of these presents, the only telescope that Mary got was from Joan. Joan or other people may have given people other than Mary telescopes.

b. Joan gave **Mary**$_F$ the only telescope.
≈ Joan gave some people presents. Out of these presents, the only telescope that Joan gave was to Mary. Mary or other people may have received telescopes from people other than Joan.

The focus-sensitive readings disappear if there is a relative clause associated with the NP that *only* modifies. There do not seem to be two truth-conditionally different readings depending upon where the focus is placed.

(118) a. **Joan**$_F$ gave Mary the only telescope [to be built in the 9th century].
≈ It was Joan who gave Mary the only telescope [to be built in the 9th century].

b. Joan gave **Mary**$_F$ the only telescope [to be built in the 9th century].
≈ It was to Mary that Joan gave the only telescope [to be built in the 9th century].

Unlike (117a, b) which are truth-conditionally different, (118) differ only in their presuppositions. Paraphrases for the missing readings are provided below in (119).

(119) a. **Joan**$_F$ gave Mary the only telescope [to be built in the 9th century].
≠ Some people gave Mary presents. Out of these presents, the only 9th century telescope that Mary got was from Joan. Joan or other people may have given people other than Mary 9th century telescopes.

b. Joan gave **Mary**$_F$ the only telescope [to be built in the 9th century].
≠ Joan gave some people presents. Out of these presents, the only 9th century telescope that Joan gave was to Mary. Mary or other people may have received 9th century telescopes from people other than Joan.

The facts regarding the loss of association in the presence of a non-modal infinitival relative follow from the licensing configuration proposed for the licensing of the non-modal interpretation of infinitival relative clauses and thus provides support for the proposal also. If the focus-sensitive licensor was outside the relative clause, we would expect it to be able to move and associate with focus as it does in the absence of a non-modal infinitival relative clause. The loss of association with focus in presence of a non-modal infinitival relative would not be predicted.

5.2. A further prediction

We make another prediction that in cases involving relative clauses where the superlative/ordinal/*only* does not have to be interpreted CP-internally, association with focus readings should be present. We have seen that superlatives/ordinals/*only* stop associating with focus if they are associated with a non-modal infinitival relative clause. This is explained by the assumption that the licensor of the non-modal interpretation has to be inside the relative clause at LF for it to be able to license the non-modal reading. The relative clause-internal interpretation of the superlative/ordinal/*only* is, however, in general not forced upon us. In the case of non-modal infinitival relative clause, the relative clause-internal interpretation of superlative/ordinal/*only* is the only option since relative clause-external interpretation leads to failure of licensing and hence ungrammaticality.

When the relative clause does not need to be licensed, as is the case with object infinitival relatives and finite relative clauses, the superlative/ordinal/ *only* does not have to be interpreted CP-internally. In principle, the superlative/ ordinal/ *only* does not have to even start off with the rest of the NP inside the relative clause (cf. 120).

(120) a. the longest book that John wrote
possible structure: the [$_{AP}$ longest [$_{NP}$ [$_{NP}$ book] [$_{CP}$ [$_{NP}$ *book*] that [John wrote [$_{NP}$ *book*]]]]]
LF for this structure: the [$_{AP}$ longest [$_{CP}$ [$_{NP}$ *book*]$_i$ that [John wrote t_i]]]

b. the longest book to read
possible structure: the [$_{AP}$ longest [$_{NP}$ [$_{NP}$ book] [$_{CP}$ [$_{NP}$ *book*] C^0 [PRO to read [$_{NP}$ *book*]]]]]
LF for this structure: the [$_{AP}$ longest [$_{CP}$ [$_{NP}$ *book*]$_i$ C^0 [PRO to read t_i]]]

(121) shows that association with focus readings are available with finite relative clauses and (122) does the same for object infinitival relatives.

(121) a. **Joan**$_F$ gave Mary the most expensive telescope that was built in the 9th century.
≈ Of the people that gave Mary 9th century telescopes, the telescope that Joan gave Mary was the most expensive.

b. Joan gave **Mary** the most expensive telescope that was built in the 9th century.
$\approx$ Of the people that got 9th century telescopes from Joan, the telescope that Mary got from Joan was the most expensive.

(122) a. **Joan**$_F$ assigned Mary the longest book to read.
Of the people that assigned Mary books to read, the book assigned by Joan was the longest.

b. Joan assigned **Mary**$_F$ the longest book to read.
Of the people that Joan assigned books to read, the book assigned to Mary was the longest.

6. Interpretation of the non-modal infinitival clauses

In order to explore the temporal/aspectual properties of the Non-modal Infinitival Relative clause, I will use Ordinals as a probe. Let us therefore begin with a discussion of the semantics of Ordinals. For simplicity, I will discuss the semantics of *first*, which can then be extended to handle other ordinals.

6.1. A first semantics for *first*

We begin with the observation that *first* is not an intersective modifier. It looks at more than the extension of its complement. If the set of astronauts and the set of moonwalkers are co-extensive, it does not follow that 'the first astronaut' is the 'the first moonwalker'. This is in contrast to intersective modifiers e.g. *American* which look only at the extension of their sister node. So if the set of astronauts and the set of moonwalkers are co-extensive, it does follow that 'the American astronauts' are identical to the 'the American moonwalkers'.

Consider a simple NP with an ordinal modifier like *the first astronaut*. *first* takes the set of astronauts, sorts the set according to when the members of the set acquired the *astronaut* property i.e. entered the set and returns a set that contains the first member of the sorted set as its unique member. We can now define the semantics of *first*.

(123) $first(P) = \lambda x \lambda t.[P(x,t) \wedge \forall y[[x \neq y \wedge \exists t'.P(y,t')] \Longrightarrow F(P)(x) < F(P)(y)]]$

where

$F(P)(x) = \iota \lambda t.[P(x,t) \wedge \forall t' < t. \neg P(x,t')]$

The function F when applied to a predicate of times and individuals and an individual yields the time when that individual came to have the property P i.e. it yields the time at which the flip from $\neg P$ to P took place. The complement of *first* is a predicate of times and individuals.

Note that $first(P)$ is not a timeless property: someone has the property $first(P)$ only as long as they have the property P. In the case at hand, *first* (*astronaut*) seems to be timeless - once somebody acquires the property *first* (*astronaut*), they have it for the rest of their life. This is presumably because once a person goes out into outer space, that person acquires forever the *astronaut* property. Since the past cannot be changed, once someone acquires the *first* (*astronaut*) property, it stays with them for the rest of their life.

6.1.1. *first with possessive NPs*

We will now consider cases where there is more than one property that can vary temporally. We begin with a discussion of NPs with possessive modifiers

(124) John's first restaurant

In (124), there are two things that can vary temporally: (i) the possession relationship, and (ii) restauranthood. The most easily accessible reading is the one where it is the time of John's acquiring possession that determines what is 'John's first restaurant' i.e. the possession relationship has to be in the scope of *first* However, this is not the only reading available. 'John's first restaurant' could pick out the first thing that belongs to John to become a restauraunt. This case is similar to the case of *former* with possessive modifiers discussed by Larson & Cho (1998). In the following section, we briefly go over Larson & Cho's analysis and then present our analysis for *first*.

6.1.2. Digression: Larson & Cho (1998)'s analysis of former

Larson & Cho 1998 propose that (125) has the two readings indicated in (125a) and (125b).

(125) This is John's former restaurant.

a. CN-Modification: This is (now) John's & formerly, this was a restaurant. (CN = Common Noun)
LF: John's former [restaurant]

b. P-Modification: Formerly, this was John's & (now) this is a restaurant. (P = Possession)
LF: former [John's restauraunt]

In the structure indicated for the P-modification reading in (125b), *former* applies to the property formed by the conjunction of the possession by John property and the restaurant property. This is possible if, as Larson & Cho assume following Kayne (1993), the possessive starts off lower in the NP and can be interpreted reconstructed inside the NP. Therefore at LF, the possessive relationship can be in the scope of *former*.

6.1.3. Back from the digression

We will assume, as Larson & Cho do, that the prenominal genitive gets to its surface structure position via movement from inside the NP and can be reconstructed inside the NP. The definiteness is done by a covert definiteness marker at the top. We will also assume that $first$ modifies the conjunction of the possession relation and the restauranthood property. Let us call this the Conjunction Modification analysis. (cf. 126).

(126) restaurant (of) John's = $\lambda x \lambda t[restaurant(x,t) \wedge Poss(x,j,t)]$
Def^0 $first$ restaurant (of) John's =
$$\iota first(\lambda x \lambda t[restaurant(x,t) \wedge Poss(x,j,t)])$$

In the above representation, the property that $first$ applies to is
$$\lambda x \lambda t[restaurant(x,t) \wedge Poss(x,j,t)]$$
which is a conjunction of two properties, the possession by John property and the restaurant property. Under the Conjunction Modification analysis, $first$ has access to the temporal parameters of both *restaurant* and the Possession property. We now go over some features that fall out of the analysis in (126).

In the conjunction modification analysis, *first* is not more closely related to one property or the other. It applies only to the conjunction of the two properties and either property can determine what 'the first P & Q' is. Cases where it is the possession relationship (= P) that determines what 'the first P & Q' is are easy to come up with e.g. John's first restaurant is the first restaurant that John bought. Cases where it is the restaurant property (= Q) that determines what 'the first P & Q' is require more contextual work but are available. This is shown in scenario 1.

(127) Scenario 1:
In 1970, John owned many delis $A, \ldots, Q$. In 1971, he started to convert them into restaurants. Now all the delis are restaurants. Deli *K* was the first one to be converted into a restaurant.

Here, it is the time of acquiring the property of restauranthood that determines what counts as *John's first restaurant*. *K* is *John's first restaurant* and it is the restauranthood of *K* that undergoes a change of state and not its possession relationship with John. Further, suppose of the delis that were converted into restaurants, John had acquired Deli *B* first. This does not effect what *John's first restaurant* is. In other words, John's first restaurant is not the first thing that John bought that is a restaurant now (= *B*).

The CN-Modifying reading is structurally available but odd perhaps due to the uniqueness of *first/only/-est* N. I propose the following felicity condition on possessive constructions of the form 'XP's YP': the denotation of YP is not asserted to be a singleton set. In cases like *Laura's Dan*, this condition forces a non-proper name interpretation on the proper name *Dan*. However, nominals of the form *first/only/-est* N have uniqueness as part of their meaning and cannot be coerced in the way proper names like *Dan* could be. Consequently, the CN-modification reading which should be available in examples like *Edison's first telephone* (≈ the x such that x is the first telephone in the world and Edison owns x) is quite odd.

An important aspect of the semantics of *first* and one that is often obscured is that $first(P)$ returns a non-timeless property: This can be seen in scenario 2.[37]

(128) Scenario 2:
The restaurants in Scenario 1 prove to be unprofitable and John converts them into laundromats. In particular restaurant *K* becomes a laundromat.

Now we find the following contrast:

(129) a. #This laundromat/#This/#K/#Laundromat K is John's first restaurant.
(judged false/infelicitous)

b. This laundromat/This/K/Laundromat K was/used to be John's first restaurant.

The fact that (129a) is judged as false/infelicitous shows that *K*, which is no longer a restaurant, does not qualify as *John's first restaurant* anymore. In other words, the property of being *John's first restaurant* is not a timeless property. *K* had it at one point but does not have it anymore. (129b) is acceptable because at a certain point in the past, *K* did have the property of being *John's first restaurant.*

The contrast between (129a) and (129b) follows from the semantics that we have proposed for *first*. In (129a, b), *first* applies to a property of time and individuals P (= $\lambda t \lambda x[restaurant(x,t) \wedge Poss(j,x)]$), and returns another property of times and individuals $first(P)$. According to the semantics provided for *first* in (123), x at time t has the property $first(P)$ only if x at time t has the property P. For (129a) to be judged as true, $first(P)(K,t_{Now})$ would have to be true. Since $P(K,t_{Now})$ is false, $first(P)(K,t_{Now})$ cannot be true. Hence (129a) is judged as false. (129b) is judged as true because there is a point in the past, call it t_{Past} s.t. $first(P)(K,t_{Past})$ is true (see Scenario 1 in 127). At t_{Past}, K was a restaurant.

6.2. *first* with non-modal infinitival relatives

We will now use the semantics developed for *first* in the previous sections with cases of non-modal infinitival relatives like (130).

(130) The first person to walk on the moon was American.
LF of subject NP: the [first [person to walk on the moon]]

Let us assume right now that there is no tense information in the infinitival clause, an assumption that will be revised later. Interpreting the LF indicated in (130), we get (131).

(131) a. $[\![the\ first\ person\ to\ walk\ on\ the\ moon]\!] =$
$\iota \lambda x((first(P))(x,t))$ (tense variable t is left unbound)[38]
$P = \lambda t \lambda x.[person(x,t) \wedge walk(x,m,t)]$

b. $[\![the\ first\ person\ to\ walk\ on\ the\ moon\ was\ american]\!] =$
$\exists t.[t < t_{NOW} \wedge american(\iota((first(P))(t)),t)]$
(unbound tense variable *t* is bound by matrix tense)

6.3. Temporal simultaneity between 'head' NP and relative clause

We had observed earlier that in a non-modal infinitival relative clause the head NP and the infinitival clause have to hold simultaneously. This can be seen in (132), where we have the intuition that in order for someone to qualify as *the first 80 yr. old to walk on the moon*, they must have been 80 yrs. old when they walked on the moon.

(132) The first 80 yr. old (person) [$_{RC}$ to walk on the moon] was American.

So Neil Armstrong will not become *the first 80 yr. old to walk on the moon* simply by turning 80.

The semantics proposed for *first* combined with independently motivated assumptions relating to the temporal interpretation of NPs yield the simultaneity between the interpretation of the head NP and the infinitival clause.

Following Musan (1997), I assume that the time variables of 'weak' NPs cannot be set independently of matrix tense. It had been noted by Enç (1986) that noun phrases are able to set their temporal variable independently of the matrix tense. Examples like (133) are used to demonstrate this point.

(133) Every fugitive is now in jail.

If the temporal variable of the subject *every fugitive* was obligatorily bound by the matrix present tense, (133) would only have a contradictory reading. However, (133) has a perfectly sensible reading where it means that every person who was a fugitive at some time before now is now in jail. This demonstrates that *every fugitive* is able to set its own temporal variable. Musan (1997) pointed out that Enç's generalization covers only definites and strong NPs. She notes that the temporal variable in weak NPs cannot be set independently (cf. 134).

(134) a. There were many homeless people at the rally.

b. Many of the homeless people were at the rally.

In (134a), where only a weak interpretation of *many homeless people* is available, the relevant people have to have been homeless at the time of the rally. In contrast, if a strong interpretation is forced (as in 134b), the relevant people may have been homeless at some time before the rally.

In our proposal, the reconstructed NP is always weak because it lacks the quantificational DP layer. From Musan's observation, we know that the temporal variable of a weak NP cannot be set independent of its syntactic environment. Therefore, the temporal interpretation of the reconstructed NP cannot be independent of its syntactic environment. I assume that the weak NP forms a predicate of times and individuals and combines via Predicate Modification with the infinitival clause, which is also a predicate of times and individuals. As a result of Predicate Modification, the temporal variable of the reconstructed NP is identified with the temporal variable of the infinitival clause. Cases like (135=132), which involve a temporally variable 'head' NP, are handled properly:

(135) The first 80 yr. old (person) [$_{RC}$ to walk on the moon] was American.
$\exists t.[t < t_{NOW} \wedge american(\iota(first(P))(t),t)]$
$P = \lambda t \lambda x[80yr.old(x,t) \wedge walk(x,m,t)]$

In the logical representation in (135), the head NP and the infinitival predicate have to hold at the same point of time/time interval. This captures our intuition that (135) does not mean the first person to obtain both properties. As we noted earlier, Neil Armstrong will not become 'the first 80yr. old to walk on the moon' when he turns 80.

6.3.1. *Simultaneity effects with finite relative clauses and* first

The simultaneity requirement between the head NP and the infinitival clause that we have seen in non-modal infinitival relatives holds for finite relative clauses with *first* also (cf. 136).

(136) The first 80 yr. old (person) who walked on the moon was American.

As with the Non-Modal Infinitival Relatives, the 'first 80 yr. old (person) who walked on the moon' has to have been 80 yrs. old when that person walked on the moon. This is unexpected under a head-external analysis of relative clauses.[39] In fact, the head-external analysis seems to predict non-simultaneity. To see why, assume that the head NP and the relative clause are both interpreted as predicates of times and individuals.

But before we do this, we need to make some decisions regarding how to formally represent the past tense in the relative clause. (137) shows some of the options that have been discussed in the literature on tense.

(137) John ate the apple.
Referential Past: $ate(j,a,t_{Pst}) \wedge t_{Pst} < t_{Now}$
Indexical Past: $\exists t[ate(j,a,t) \wedge t < t_{Now}]$
Relative Past: $\lambda t \exists t'[ate(j,a,t') \wedge t' < t]$

The referential and indexical past options return propositions and not predicates of times. In order to permit the finite past tense relative clause to be a predicate of times and individuals, we temporarily assume that the past tense in the relative clause is interpreted as a relative past. In sect. 6.3.2, the empirical consequences of adopting the referential/indexical past formalization will be compared with the consequences of adopting the relative past formalization.

Having adopted the relative past formalization for the past tense in the relative clause, we get the logical representation in (138).

(138) $[\![80\ yr.\ old]\!] = \lambda t \lambda x[80yrold(x,t)]$
$[\![who\ walked\ on\ the\ moon]\!] =$
$\lambda x \lambda t \exists t'[t' < t \wedge walk(x,m,t')]$
Applying Predicate Modification
$[\![80\ yr.\ old\ who\ walked\ on\ the\ moon]\!] =$
$\lambda x \lambda t[80yrold(x,t) \wedge$
$\exists t'[t' < t \wedge walk(x,m,t')]] = P$
$[\![the\ first\ 80\ yr.\ old\ who\ walked\ on\ the\ moon]\!] = \iota((first(P))(t))$
(the temporal variable t is left free)

What the logical representation in (138) gives us is the first person to acquire to acquire the two properties of being 80 yrs. old and of having walked on the moon. This is precisely what we do not want.[40]

Under the Raising analysis of relative clauses, we can provide a simple analysis of the simultaneity facts with finite relative clauses and *first*. Assume that the 'head NP' (*80 yr. old*) is interpreted in its trace position inside the relative clause. We have two options for where *first* is interpreted - it can be interpreted outside the relative clause or inside it. These two options correspond to the two distinct base structures for the relative clause shown in (139). The string in (139) is structurally ambiguous between the structures indicated in (139a) and (139b).

(139) The first 80 yr. old (person) who walked on the moon

a. Structure 1: the [first [$_{NP}$[$_{NP}$ 80 yr. old person] [$_{RC}$ who walked on the moon]]]
Only NP can reconstruct into the relative clause, *first* cannot.

b. Structure 2: the [[$_{NP}$ first 80 yr. old person] [$_{RC}$ who walked on the moon]]
first can also reconstruct into the relative clause.

The NP *80 yr. old person* is interpreted relative clause-internally; if structure 1 is interpreted, *first* is interpreted outside the relative clause (cf. 140a).[41]

(140) a. *first* outside the RC, above the RC Past Tense
the [*first* [$_{CP}$ λx [Pst [[80 yr.old,x] [walk on the moon]]]]]
'the first to have had property *P*'

$$\iota((first(\lambda t \lambda x \exists t''[t'' < t \wedge 80yr.old(x,t'') \wedge walk(x,m,t'')]))(T_{context}))$$
$$P = \lambda x \lambda t \exists t''[t'' < t \wedge 80yr.old(x,t'') \wedge walk(x,m,t'')]$$

b. *first* inside the RC, below the Past Tense
the [$_{CP}$ λx [$_{IP}$ Pst [*first* [80 yr.old [walk on the moon]]]]]
'*x* s.t. at some point in the past *x* was the first to have property *P*'

$$\iota \lambda x[(\lambda t.\exists t'[t' < t \wedge first(\lambda t''.[80yr.old(x,t'') \wedge walk(x,m,t'')])(t')])(T_{context})]$$
$$P = \lambda t''.[80yr.old(x,t'') \wedge walk(x,m,t'')]$$

If structure 2 is interpreted, *first* is interpreted inside the relative clause. The question is still left open whether *first* will be interpreted above the RC past tense or below. If it is interpreted above, we get the logical representation in (140a) and if it is interpreted below, we get the logical representation in (140b). If the order of composition is (Tense (first [NP RC])) (=140b), *first* binds the temporal variables of NP and RC simultaneously and its output is bound by Tense. If the order of composition is (first (Tense [NP RC])) (=140a), the temporal variable of the NP and the RC are simultaneously bound by Tense and its output is bound by *first*. Either way, the temporal variables of the NP and the RC are identified yielding simultaneity. The two difference structures in (140) seem to yield logical representations which seem truth-conditionally identical for the most part. They only come apart in fairly artificial contexts where the intuitions are obscure.

6.3.2. *On why there is reconstruction with* first

Central to the account of simultaneity in the interpretation of the head NP and the relative clause[42] is the reconstruction of the head NP into the relative clause. In the case of non-modal infinitival relatives, reconstruction was forced because otherwise the non-modal reading would not be licensed. The question therefore arises why the reconstruction of the head NP is forced in finite relative clauses.

It would be convenient if the simultaneity facts followed from a general requirement for the head NP of a relative clause to be interpreted relative clause internally. Then the simultaneity requirement would not be just a fact about *first*. (141a, b) show that this is not correct.

(141) a. I know an 80 yr. old (person) who (once) walked on the moon.

b. Rashmi has invited the 80 yr. old (person) who (once) walked on the moon.

It seems clear in (141a, b) that the 80 yr. old-ness and the walking on the moon do not have to coincide. They can coincide but unlike the cases with *first* they do not have to. The simultaneity requirement does not hold for all relative clauses.

We discuss now an approach from which the reconstruction requirement follows. In the previous discussion, we assumed that the past tense should be formalized as a relative past. Let us now assume that this assumption was incorrect and that instead past tense should be formalized as a referential/indexical past.[43] If this is the case, the reconstruction of the NP is forced upon us by the type-requirements of *first*. *first* needs a predicate of times and individuals (2-argument analysis: semantic type $< T, < e,t >>$) or a predicate of times (1-argument analysis: semantic type $< T,t >$). In its surface position (assume the structure is (first (NP RC))), RC could only vacuously be made a predicate of times and individuals. So in order for *first* to have a predicate of the appropriate semantic type, we would need to reconstruct it into the relative clause. *first* can only be reconstructed if it is part of the NP head and if *first* is reconstructed, the NP head is too. Thus the reconstruction requirement on the head NP follows from the semantic properties of *first*.

This type-driven approach makes certain empirical predictions which seem to be on the right track. Under the type-driven approach, reconstruction is forced by the type requirements of *first*. Superlatives and *only* do not have the same kind of type requirement. Hence we would not be surprised to find no obligatory reconstruction and hence no simultaneity requirement with them. This seems to be the case with (142).

(142) a. Olafur introduced Miguel to the tallest 80 yr. old who walked on the moon.

b. Olafur introduced Miguel to the only 80 yr. old who walked on the moon.

In (142a, b), the person introduced to Miguel does not have to have been 80 yrs. old when they walked on the moon.

The type-driven approach connects the simultaneity requirement to the location of *first* - when the simultaneity requirement holds *first* is interpreted internal to the relative clause. We know the following from the discussion of the licensing of non-modal infinitival relatives: when ordinals, superlatives, and *only* were forced to be inside the relative clause at LF for licensing reasons, they were unable to associate with matrix focus. Now consider (143).

(143) a. Olafur introduced **Miguel**$_F$ to the first 80 yr. old who walked on the moon.
$\approx$ Olafur introduced many people to 80 yr. olds who walked on the moon. The first introduction was to Miguel. The moonwalkers did not have to have been 80 when they walked on the moon.

b. **Olafur**$_F$ introduced Miguel to the first 80 yr. old who walked on the moon.
$\approx$ Many people introduced Miguel to 80 yr. olds who walked on the moon. The first introduction was by Olafur. The moonwalkers did not have to have been 80 when they walked on the moon.

Association with focus does seem to be possible as can be seen by the truth-conditionally distinct paraphrases of (143a, b). However, when there is association with focus, two other things happen: the simultaneity effect disappears and *first* only looks at the temporal variable of *introduce* and not at the conjunction of *introduction*, *80 yr. old*, and *walked on the moon*. This set of facts receives a natural explanation under the type-driven approach. The simultaneity effect disappears when there is association with focus because the *first* at LF is associated with *introduce*. There is no *first* in the NP and so we essentially have the structure in (141). As in (141), nothing forces the reconstruction of the head NP and therefore we do find a simultaneity requirement. Association with focus is not ruled out since *first* does not have to be inside the relative clause at LF. This is unlike the case with non-modal infinitival relatives where the *first* has to be inside the relative clause for licensing reasons and hence the association with focus reading is ruled out altogether. When *first* associates with focus, it is only able to look at the temporal variable of *introduce* and not at the temporal variables of the relative clause because (by assumption) in order to access the latter, it would have to be interpreted internal to the relative clause.

So empirically the type-driven approach for forcing reconstruction seems promising. However, there are problems of a conceptual nature that this approach faces. The approach crucially relies on the unavailability of a mechanism that allows us to interpret a simple past tense as a relative past tense.

(144) John ate the apple.
Referential Past: $ate(j,a,t_{Pst}) \wedge t_{Pst} < t_{Now}$
Indexical Past: $\exists t[ate(j,a,t) \wedge t < t_{Now}]$
Relative Past: $\lambda t \exists t'[ate(j,a,t') \wedge t' < t]$

We have to assume that only the mechanisms of indexical past/existential past are available and not that of Relative past (cf. 144). However, it is unclear why the relevant mechanism should be unavailable. Unless there are arguments to support such a move, and we have not provided any here, we do not have a proper explanation.

6.4. A prediction: locus of change of state is undetermined

In the proposed analysis, *first* takes a conjunction of two predicates of times as an argument. This can be seen in (145), where the argument of *first* is a conjunction of $\lambda t.person(\mathbf{x},t)$ and $\lambda t.walk(\mathbf{x},m,t)$.

(145) a. $[\![$*the first person to walk on the moon*$]\!] =$
$\iota((first(P))(t))$ (tense variable *t* is left unbound)
$P = \lambda x \lambda t.[person(x,t) \wedge walk(x,m,t)]$

b. $[\![$*the first person to walk on the moon was american*$]\!] =$
$\exists t.[t < t_{NOW} \wedge american((\iota(\lambda x[(first(P))(x,t)])),t)]$
(unbound tense variable *t* is bound by matrix tense)

first is not more closely associated with one property or the other.[44] *first* applies to a predicate P which is a predicate of times and individuals. It picks out the first x that comes to have P i.e. it looks at a change from $\neg P$ to P. We predict that since *first* does not have special with either the head NP or the infinitival clause, the source of the change from $\neg P$ to P i.e. the locus of the change of state is not pre-determined to be the infinitival clause (or for that matter, from the head). It could be the infinitival clause or the head NP. This prediction is met (cf. 146).

(146) a. Junko Tabei was the first woman to climb Mt. Everest.
Locus of Change of State: 'climb Mt. Everest' (infinitival clause)

b. Kennedy was the first president to come from a Catholic family.
Locus of Change of State: 'President' (head noun)

c. The first astronaut to be blond was from Wisconsin.

Reading 1: The first astronaut who became blond was from Wisconsin
Locus of Change of State: 'be blond' (infinitival clause)

Reading 2: The first blond person to become an astronaut was from Wisconsin
Locus of Change of State: 'astronaut' (head noun)

In (146a), the locus of the change of state is the infinitival clause. The locus of the change of state in (146b) lies in the head NP. Kennedy's coming from a Catholic family is not what changes. Finally, with (146c) the locus of the change of state can be either the head NP or the infinitival clause. If the locus of the change of state is the infinitival clause, the reading can be paraphrased by 'the first astronaut to become blond (i.e. have his hair dyed blond) was from Wisconsin' and if the locus of the change of state is the head NP, 'the first astronaut to be blond' picks out 'the first blond person to become an astronaut' .

6.5. Temporal properties of the infinitival clause

Recall that by the definition of *first* (repeated in 147), if $first(P)(x,t)$ is true, then $P(x,t)$ is also true.

(147) $first(P) = \lambda x \lambda t.[P(x,t) \wedge \forall y[[x \neq y \wedge \exists t'.P(y,t')] \Longrightarrow F(P)(x) < F(P)(y)]]$

where

$F(P)(x) = \iota \lambda t.[P(x,t) \wedge \forall t' < t.\neg P(x,t')]$

In the case of non-modal infinitival relatives, *P* is created by predicate modification of two predicates, one corresponding to the head NP and the other corresponding to the infinitival clause. So far, we have made the simplifying assumption that there is no aspectual information in the infinitival clause. So *to walk on the moon* was directly interpreted as $\lambda t \lambda x[walk(x,m,t)]$.

We will address two issues in this section: (i) when is $first(P)$ evaluated, and (ii) is the assumption that there is no aspectual information in the infinitival correct? If not, what kind of aspectual information is present in the infinitival clause?

We will argue that in non-predicative positions $first(P)$ can be evaluated at any contextually salient present or past time. However, in predicative positions $first(P)$ has to be evaluated at a time determined by the matrix tense. It will be shown that the simplifying assumption that there is no aspectual information in the infinitival clause will have to be given up. Depending upon whether the infinitival clause is stative or non-stative, the predicate that it will denote will have different temporal properties.

6.5.1. When is $first(P)$ evaluated?

The semantics for *first* that we have proposed has the result of applying *first* to *P*, a property of times and individuals, also be a predicate of times and individuals i.e. its semantic type is $<< T,< e,t >>,< T,< e,t >>>$. *The first P*, however, seems to pick out an individual. The definite article *the* gives us an individual from a predicate so in order to get an individual from $< T,< e,t >>$, we need to satisfy the *T* argument. We can do this in two ways (cf. 148a, b):

(148) the first P

a. $\iota((first(P))(T))$

b. $(\iota^T(first(P)))(T)$

(assume an intensionalised ι, where $\iota^T(P) = \lambda T[\iota(P(T))]$

We noted earlier Musan (1997)'s observation that definite/strong NPs are able to set their temporal variable independently of matrix tense while weak/ predicative NPs seem to be dependent on the tense of the clause in which they occur. Her observation suggests that free setting of temporal variables is only possible in association with a strong determiner. I will therefore adopt (148b) as the way in which the temporal variable of $first(P)$ is set. Doing so will ensure that we are able to evaluate $first(P)$ at a time independent of the matrix tense only in association with a strong determiner. The use of unbound temporal variable will be restricted to the intensionalized versions of strong determiners.

How is the temporal variable T in (148) set? Our proposal, which incorporates Musan's observation, gives us the correct results with examples like (149).

(149) The first person to walk on the moon was American.
$\exists t'[t' < t_{Now} \wedge american(\iota((first(P))(t_{Context})), t')]$
$P = \lambda x \lambda t[person(x,t) \wedge walk(x,m,t)]$

$t_{Context}$ is a contextually salient time: in most cases, it is either t_{Now} or some point of time in the past. It does not seem possible for $t_{Context}$ to be in the future. Assume that $t_{Context} < t_{Now}$ in (149). The prediction then is that the walk on the moon and the american-ness are unordered with respect to each other - the walk could have taken place before the person concerned became American, while that person was American, or after that person gave up American citizenship.

The same approach is in principle extensible to cases like (150), where the relevant NP occurs in a postcopular position.

(150) Neil Armstrong is the first person [$_{RC}$ to walk on the moon].
$BE(NA, \iota\lambda x(first(P)(x)(t_{Context})), t_{Now})$
$P = \lambda x \lambda t.[person(\mathbf{x},t) \wedge walk(\mathbf{x},m,t)]$

However, if we believe that the postcopular NP in (150) should be treated as a predicate nominal and that the temporal variable of predicative NPs is dependent upon matrix tense, we do not have the logical representation in (150). Instead, we generate the logical representation in (151).[45]

(151) Neil Armstrong is the first person [$_{RC}$ to walk on the moon].
$(((first(P))(NA))(t_{Now}))$
$P = \lambda x \lambda t.[person(\mathbf{x},t) \wedge walk(\mathbf{x},m,t)]$

This logical representation represents a very implausible reading of (151), where the walking on the moon and *NOW* have to overlap.

At this point, we could give up on the predicative analysis as the only analysis of the postcopular NP in (151). The equative analysis in (150) gives the right results. However, if we did this we would lose our explanation for the oddness of (129a, repeated as 152a).

(152) Scenario 2:
The restaurants that John owns prove to be unprofitable and John converts them into laundromats. In particular restaurant *K* becomes a laundromat.

a. #This laundromat/#This/#K/#Laundromat K is John's first restaurant.

b. This laundromat/This/K/Laundromat K was/used to be John's first restaurant.

John's first restaurant does not hold of *K* in the present since *K* is no longer a restaurant, which is why (152a) is judged as odd. *John's first restaurant* does hold of *K* in the past, explaining why (129b) is ok. If it was possible for the postcopular NP in (152a) to set its temporal variable independently of matrix tense, (152a) would not be odd. In order to retain the explanation for the contrast between (152a and b), we propose a way that makes feasible again the line that free temporal variable setting is unavailable to the postcopular NP.

6.5.2. *Aspectual characterization of the infinitival clause*

In (151), we have made the assumption that there is no aspectual information in the infinitival clause. This means that '*to walk on the moon*' is interpreted as $\lambda x \lambda t[walk(x,m,t)]$. $first(P)$ is evaluated at t_{Now} and by the semantics of $first$, if $first(P)(x,t)$ is true, $P(x,t)$ is also true. Therefore, (151) only yields the implausible reading where the walking on the moon and t_{Now} have to overlap. We will show that once we assume that there is aspectual information in the infinitival clause, the problem at hand disappears.

Before we go on to articulating the nature of the aspectual information, let us review some of the features of the current analysis that we want to hold on to. We will call the first feature the **Head NP** requirement, by which in a predicative position the 'head NP' holds at a time determined by matrix tense. In the present tense, this means that the 'head NP' should hold at t_{Now}.

(153) # Shirley Temple is the first child to get an Oscar.

(153) is odd because Shirley Temple is no longer a child.

The second feature, we will call the **Simultaneity** requirement:. It states that the head NP has to hold at the point the infinitival clause (or the event corresponding to the infinitival clause) holds.

(154) The first 80 yr. old (person) to walk on the moon will visit our school next year.

(154) requires that the person in question was 80 yrs. old when they walked on the moon.

6.5.3. *A covert perfect?*

One possibility is that there is a covert perfect in the infinitival clause. In other words, (151) is for interpretive purposes like (155).

(155) Neil Armstrong is the first person [$_{RC}$ to have walked on the moon].
$(first(P))(NA, t_{Now})$
$P = \lambda x \lambda t \exists t'[t' < t \wedge person(\mathbf{x}, t') \wedge walk(\mathbf{x}, m, t')]$
(this is a oversimplified analysis of the perfect; a more satisfactory treatment is presented later.)

One piece of evidence that supports the idea that there is a covert perfect in the infinitival clause is that adverbs that otherwise appear only with perfects seem to be acceptable with non-modal infinitival relatives. McCoard (1978) (also see Dowty (1979)) gives a list of adverbs that distinguish between the simple past and the perfect.

(156) a. *Occur with simple past but not with present perfect*:
long ago, five years ago, once, the other day, those days, last night, in 1900, at 3:00, no longer

b. *Occur with present perfect but not with simple past*:
at present, up till now, so far, as yet, during these last five years, herewith, lately, since the war, before now.

The behavior of one 'past' adverb and one 'perfect' adverb is exemplified in (157).

(157) a. *yesterday*: ok with past, simple or perfect, not ok with present perfect

i. John left yesterday/John was sick yesterday.

ii. John had left yesterday/John had been sick yesterday.

iii. *John has left yesterday/*John has been sick yesterday.

b. *since 1971*: ok with perfect, present or past, not ok with simple past

i. *John lives in Boston since 1971.

ii. John had lived in Boston since 1971.

iii. John has lived in Boston since 1971.

We find that perfect adverbs are acceptable with Non-Modal Infinitival Relatives.

(158) Perfect adverbials are acceptable

a. John is the first/only/tallest American [$_{RC}$ to visit us lately].

b. John is the first/only/tallest American [$_{RC}$ to visit us since the war].

Past adverbials seem to only be acceptable if there is matrix past tense.

(159) Past adverbials are acceptable if there is a past in the higher predicate

a. # John is the first/only/tallest American [$_{RC}$ to visit us yesterday].

b. John was the first/only/tallest American [$_{RC}$ to visit us yesterday].

If 'perfect' adverbs diagnose the presence of a perfect, we can conclude that a covert perfect can appear in the semantics of the infinitival clauses that appear with ordinals (e.g. *first*), superlatives, and *only*.

6.5.4. *A problem with assuming a covert perfect and a solution*

Let us see how things work out if we assume a covert perfect.

(160) Luis is the first 80 yr. old to walk on the moon.
LF: Luis is the [first [[80 yr. old] to *have* walk on the moon]]
Logical representation: $first(P)(l,t_{Now})$
$P = \lambda x \lambda t[80(x,t) \wedge \exists t'[t' < t \wedge walk(x,.m.t)]]$

Since $first(P)(l,t_{Now})$ is true, $P(l,t_{Now})$ is also true. Therefore it follows that Luis has to be 80 yrs. old now. Thus the **Head NP** requirement is met. However according to the logical representation, Luis does not have to be 80 yrs. old when he walked on the moon. This does not match our intuitions about (160); The **Simultaneity** requirement is not met.

In order to meet the Simultaneity Requirement, which holds in (151), the head NP has to be interpreted below the *have*.

(161) Luis is the first 80 yr. old to walk on the moon.
LF: Luis is the [first [*have* [[80 yr. old] walk on the moon]]]
Logical representation: $first(P)(l,t_{Now})$
$P = \lambda x \lambda t \exists t'[t' < t \wedge 80(x,t') \wedge walk(x,.m.t')]]$

The logical representation in (161) requires Luis to be 80 yrs. old when he walked on the moon. However, it does not require Luis to be 80 yrs. old now. Now the **Simultaneity** requirement is met but not the **Head NP** requirement. We seem to be in a bind - to get simultaneity we need to reconstruct but once we reconstruct we lose the **Head NP** requirement.

The copy theory of movement provides a way to rescue the analysis with the perfect from this bind. It is sometimes assumed that given a chain CH = (XP,...,XP), the different elements that constitute XP are interpreted only once. This point is illustrated in (162).

(162) I wonder [[which picture] John likes [*which picture*]]

a. LF1: I wonder [[which *x*] John likes [*x*,picture]]

b. LF2: I wonder [[which [*x*,picture]] John likes *x*]

c. LF3: I wonder [[which [*x*,picture]] John likes [*x*,picture]]

Under this conception LF1 and LF2, but not LF3, are considered to be potential LFs for (162) because LF3 involves interpretation of the same material (*picture*) twice. However, it is not clear what this tacit assumption follows from. It does not follow from the idea that LFs are derived by deleting (enough) material from copies to yield an interpretable object. There does not seem to be any prima facie problem in interpreting LF3 so it is not clear why LF3 should be blocked. I will assume that, in general, as long as the resulting structure is interpretable, multiple copies of a chain can be interpreted. Note also that the evidence from the reconstruction literature (Romero 1997, Fox 1999, among others) only shows that a lower copy *must* be present. It does not rule out the presence of a higher copy at LF. Recent work by Fox (2001,

2002) and Bhatt and Pancheva (2004) makes crucial use of the interpretation of multiple copies.

In the analysis for relative clauses assumed here, the head NP is related via movement to a relative-clause internal position. Let us assume that the 'head NP' of a relative clause is always interpreted in its relative-clause external position. Why this should be so is not obvious to me - one explanation could be that if we delete the head NP altogether, D^o would directly have to take a *CP* complement. Presumably whatever motivates the movement of the head NP out of the relative clause also motivates the retention of (at least the head of) the head NP. Now let us re-examine the example in (160 = 163), which was problematic for the covert perfect analysis. For simplicity, let us assume that *first* is only interpreted inside the relative clause. As before, we assume that the head NP has to be reconstructed under the covert perfect.

(163) Luis is the first 80 yr. old to walk on the moon.
LF: Luis is the [[80 yr. old] [first [*have* [[80 yr. old] walk on the moon]]]]
Logical representation: $first(P)(l,t_{Now}) \wedge 80(l,t_{Now})$
$P = \lambda x \lambda t \exists t'[t' < t \wedge 80(x,t') \wedge walk(x,.m.t')]]$

Now both the **Head NP** and the **Simultaneity** requirements are met. In addition, due to the perfect, the walking does not have to be taking place right now.

If we interpret the $first$ both in the head-external position and a relative-clause internal position, the resulting logical representation (in 164) seems to have the same truth-conditions as (163).

(164) Luis is the first 80 yr. old to walk on the moon.
LF: Luis is the [first [[80 yr. old] [first [*have* [[80 yr. old] walk on the moon]]]]]
Logical representation: $first(P)(l,t_{Now})$
$P = \lambda x \lambda t[80(x,t) \wedge first(P')(x,t)]$
$P' = \lambda x \lambda t \exists t'[t' < t \wedge 80(x,t') \wedge walk(x,.m.t')]$

If $first$ can be interpreted outside the relative clause, do we then lose our explanation of the loss of association with focus effects in the presence of non-modal infinitival relatives (cf. sect. 5)? The answer is not necessarily. To see why let us first consider what the two possible structures corresponding to a relative clause with an adjective are (cf. 165).

(165) the shaggy dog that Olafur likes

a. Structure 1, A^0 is RC-external
the [$_{AP}$ shaggy [$_{NP}$ [$_{NP}$ dog] [$_{CP}$ that Olafur likes *dog*]]]

b. Structure 2, A^0 starts with the NP inside the RC
the [$_{AP}$ [$_{AP}$ shaggy dog] [$_{CP}$ that Olafur likes *shaggy dog*]]]

In the case of non-modal infinitival relatives, we need reconstruction for licensing and hence only Structure 2 is an option. The question then is whether a *first* that starts off relative clause-internally and is implicated in licensing the non-modal reading can then move out of its DP. Presumably locality constraints of the sort that block movement out of relative clause islands conspire to block such a movement. If this is the case then Structure 2 does not provide an appropriate configuration for ordinals and superlatives to associate with matrix focus. [46]

6.5.5. A difference between the covert and the overt perfect

We were able to explain the first challenge we encountered to the proposal that non-modal infinitival relatives involve a covert perfect. But there is a more serious challenge that comes from the fact that the covert perfect that we have postulated does not pattern with the overt perfect.[47]

(166) a. In two minutes time, Luis will become the first 80 yr. old to have walked on the moon.
Context: Luis walked on the moon when he was in his twenties. In two minutes, Luis will complete 80 yrs. Then he will become the first moonwalker to reach the age of 80.

b. Luis is the first 80 yr. old to have walked on the moon.

In both (166a, b), Luis is not required to have been 80 yrs. old when he walked on the moon i.e. the simultaneity requirement does not hold in the presence of an overt perfect.

The lack of parallelism between the overt perfect and its presumed covert counterpart is a serious one. Our motivation for proposing a covert perfect came from an attempt to derive certain readings and especially the acceptability of certain 'perfect' adverbs. But it is quite unclear exactly what the acceptability of 'perfect' adverbs in an infinitival clause diagnoses. Instead of diagnosing the presence of a covert element, the acceptability could be diagnosing the absence of an element/certain semantics. So given that the thrust of the adverbial data is unclear, we will take the lack of parallelism seriously and consider another approach.

6.5.6. *Perfective aspect*

Instead of assuming that there is a covert perfect in the infinitival clause, we will explore the idea that overtly aspectually unspecified non-stative verbs in English are implicitly specified for perfective aspect. The following semantics for the perfective, based on Klein (1994) and taken from Iatridou, Anagnostopoulou, and Izvorski (2001) , is assumed.

(167) Perfective: $\lambda P \lambda T \exists e[\tau(e) \subseteq T \wedge P(e)]$

$\tau(e)$ is the temporal trace of the event.

This semantics assumes that predicates start off with an event semantics and that perfective/imperfective aspect is what takes us from an event semantics to a temporal semantics. Operators like the perfect only apply after perfective/imperfective have composed with the event predication.

The subject of the infinitival relative is a predicate of times and individuals. But if temporal reference is not achieved until aspectual operators apply, it must be the case that the structure we interpret has the following LF:

(168) ..... [NP] λx Perfective [P(x)]
... [80 yr. old] λx Perfective [x walk on the moon]

type of [80 yr. old]: $< e, < T,t >>$
type of λx Perfective [x walk on the moon]: $< e, < T,t >>$

Applying this analysis to our Neil Armstrong example, we get the following results.

(169) Neil Armstrong is the first man to walk on the moon.
(He walked on the moon on July 20, 1969.)
LF: first [[man] λx Pfv [x to walk on the moon]]

$Pfv(x\ walk\ on\ the\ moon) = \lambda T \exists e[\tau(e) \subseteq T \wedge walk(x,m,e)]$
[[man] λx Pfv [x to walk on the moon]] =
$\lambda x \lambda T[man(x,T) \wedge \exists e[\tau(e) \subseteq T \wedge walk(x,m,e)]$ = P

The predicate P (= $\lambda x \lambda T[man(x,T) \wedge \exists e[\tau(e) \subseteq T \wedge walk(x,m,e)]]$) is true for every moonwalker for every interval that includes their moonwalking. So given that Neil Armstrong walked on the moon on July 20, 1969, born on August 5, 1930, and is alive in 2006, P holds of Neil Armstrong for intervals such as (1968-1970), (1969-2004), (1931-1973), and so on, but not for the intervals (1901-1929), in which he did not exist, or for intervals like (1931-1968) and (1970-2006), which do not include the moonwalking. Bringing in tense and composing with the subject, we get (170).

(170) $first(P)(NA)(t_{now}$
P (= $\lambda x \lambda T[man(x,T) \wedge \exists e[\tau(e) \subseteq T \wedge walk(x,m,e)]$

Note that t_{now} is an interval, not a point. It includes the present but can, and in this case, does stretch out into the past. [48]

At first blush, (170) seems to give us the right results - by the semantics of *first* and the homogeneity property of a predicate like *man*, Neil Armstrong has to be a man right now (the **head** requirement). And since the moonwalking is contained in the interval in which the *man* predicate holds, we also satisfy the **simultaneity** requirement. But unfortunately all is not well. To see this, let us bring in another astronaut, Buzz Aldrin. Buzz Aldrin was the second man to walk on the moon. His moonwalk was also on July 20, 1969, but followed Neil Armstrong's. As it happened, Buzz Aldrin was born on January 20, 1930, several months before Neil Armstrong. So P holds of Buzz Aldrin for an interval like (January 20, 1930, 1972) but not for Neil Armstrong. Our semantics for *first* incorrectly anoints Buzz Aldrin as the first man to walk on the moon.

The same problem arises if we have a more transitory predicate like *80 yr. old* as the head noun.

(171) $[\![first \text{ 80 } yr.\ old\ person\ to\ walk\ on\ the\ moon]\!] = first(P)$,
where $P = \lambda x \lambda T[80(x,T) \wedge \exists e[\tau(e) \subseteq T \wedge walk(x,m,e)]]$
$[\![Bill\ is\ the\ first \text{ 80 } yr.\ old\ person\ to\ walk\ on\ the\ moon]\!] =$
$first(P)(b, T_{Now})$

Now suppose both Bill and John are currently 80 years old and that they both walked on the moon in their 80th year. Also assume that Bill walked on the moon a day before John but was born a day after John. This means that Bill turned 80 years old a day after John did. The way we construct the input to *first*, the period before the moonwalking during which the moonwalker is 80 years old counts. Consequently we incorrectly predict that John, and not Bill, is the first 80 year old to walk on the moon.

The source of the problem in both cases is the fact that our definition of the perfective is allowing us to look at intervals to the left of the event time. If we focused on the left boundary of the event time, none of these problems would arise. So we propose the following modification of the definition.

(172) Perfective: $\lambda P \lambda T \exists e[\tau(e)$ is a leftmost segment of $T \wedge P(e)]$

$\tau(e)$ is the temporal trace of the event.
A is a leftmost segment of B iff $\exists C[concatenation(A,C) = B]$, where A, B, C are time intervals.

This is not a standard definition of the Perfective. It does not affect the cases where perfective yields anteriority but it does complicate the connection between intervals named by time adverbials like *yesterday* and the time interval referenced by the Perfective. In *John ate an apple yesterday*, for example, the traditional definition, but not the modified definition, can identify *yesterday* as the Perfective interval.

With this revision in hand, we return to the case of the 80 year old moonwalkers. Since we are dealing with intervals, T_{Now} can extend into the past as long as it contains NOW. The following entailments hold.

(173) $first(P)(b, T_{Now}) \Longrightarrow P(b, T_{Now})$
$P(b, T_{Now}) \Longrightarrow 80(b, T_{Now}) \wedge$
$\exists e[\tau(e)$ is a leftmost segment of $T_{Now} \wedge walk(b, m, e)]]$

80 is a stative predicate; it has the subinterval property. Therefore, Bill has to be 80 yrs. old throughout T_{Now}. $\tau(e)$ is the leftmost segment of T_{Now} and the moonwalking takes place in this segment. At the right edge is NOW. Since Bill has to be 80 yrs. old throughout T_{Now}, he has to be 80 yrs. old now (the **head NP** requirement) and 80 yrs. old at the time of the moonwalking (the **simultaneity** requirement). The fact that John became 80 years old before Bill did is not relevant anymore because pre-moonwalking periods of being 80 year old do not count.

It is important for this analysis that we are able to pick suitably extended T_{Now} intervals. In the case of *Neil Armstrong is the first man to walk on the moon*, T_{Now} must extend from NOW back to the first moonwalking on July 20, 1969. Speakers are able to do this but it requires contextual 'work'. This is the reason why while speakers accept *Neil Armstrong is the first man to walk on the moon*, they prefer (174a) and (174b).

(174) a. Neil Armstrong was the first man to walk on the moon.
$\exists T[T < T_{Now} \wedge first(P)(NA, T)]$
$P = \lambda x \lambda T[man(x, T) \wedge$
$\exists e[\tau(e)$ is a leftmost segment of $T \wedge walk(x, m, e)]]$

b. Neil Armstrong is the first man to have walked on the moon.

In both (174a) and (174b), we do not have to accommodate into T_{Now} times that we usually consider as past. Before we move on, let us look more closely at cases like (174b), which involves the perfect. We know that the cases with the perfect lack the simultaneity requirement i.e. *the first 80 year old to have walked on the moon* does not have to have been 80 years old at the time of the moonwalking. This is different from the corresponding case without the perfect: *the first 80 year old to walk on the moon*. The analysis of the perfect in Iatridou, Anagnostopoulou, and Izvorski (2001) provides the tools we need.

(175) $[\![\text{Perfect}]\!] = \lambda P \lambda T \exists T'[XN(T', T) \wedge P(T')]$
where $(XN(T, T')$ iff T is a final subinterval of T'.

The Perfect composes after the perfective/imperfective has composed with the event predicate. The case of interest is:
LF: first [NP λx Perfect (Perfective (P(x)))]

(176) Bill is the first 80 year old to have walked on the moon
LF: the (first ([80 year old] λx Perfect (Perfective (x to walk on the moon))))
$P =$ [80 year old] λx Perfect (Perfective (x to walk on the moon))
$= \lambda x \lambda T[80(x, T) \wedge$
$\exists T'[XN(T', T) \wedge$
$\exists e[\tau(e)$ is a leftmost segment of $T' \wedge walk(x, m, e)]]]$
Putting together tense:
$first(P)(b)(t_{Now})$

Bill is required to be 80 years old in T but T does not have to overlap with the moonwalking - it is a final subinterval of T', which includes the event time as a leftmost segment. One nice property of (176) is that the actual time of the moonwalking is not what matters in this case - it is the time of turning 80 years old. So in a new scenario suppose Bill only walked on the moon in 1980 but turned 80 in 2000 while Tom walked on the moon in 1970 but didn't turn 80 until 2005. Then in the reading of interest, Bill counts as the first 80 year old to have walked on the moon. This is exactly what is predicted by the logical representations in (176).

Finally if the predicate in the infinitival clause is interpreted as a stative, the predicate in the infinitival clause also has to hold at the time of evaluation (cf. 177).

(177) a. Progressive
Bill is the first 80 yr. old to be walking on the moon.

b. Generic
The Y-MP is the first Cray to use eight central processors.

c. 'Underived' Stative
This is the first wildebeest to be over 80 kilos.

Stative predicates do not involve an operator that would allow them to hold at a time before the point of evaluation and not at the point of evaluation. Consequently, in (177a-c), Bill is still walking on the moon, Y-MP uses eight central processors, and this wildebeest still weighs over 80 kilos, respectively.

6.5.7. A minimal crosslinguistic variation

Spanish permits non-modal infinitival relatives. However, it differs from English in that (178a) is judged as odd and only (178b, c) are accepted.

(178) a. *??? NA es el primer en andar por la luna*
NA is the first 'in' walk.inf on the moon
'Neil Armstrong is the first man to walk on the moon.'

b. *NA fue el primer en andar por la luna*
NA was the first 'in' walk.inf on the moon
'Neil Armstrong was the first man to walk on the moon.'

c. *NA es el primer en haber andado por la luna*
NA is the first 'in' have.inf walked on the moon
'Neil Armstrong is the first man to have walked on the moon.'

(178a) is judged odd because it requires Neil Armstrong to be on the moon now, while by world knowledge we know that he walked on the moon on July 20, 1969. In other respects, Spanish non-modal infinitival relatives pattern with English. I propose that Spanish differs from English in lacking the option of the covert perfective that makes the counterpart of (178a) acceptable in English. In the absence of the covert perfective, (178a) entails that Neil Armstrong is walking on the moon now. This is perhaps not surprising since Spanish, unlike English, has explicit marking of the perfective/imperfective distinction. More work is needed to determine whether the connection between the absence of a covert perfective in Spanish is indeed related to the explicit marking of the perfective/imperfective distinction.

6.6. Future interpretation

Non-modal Infinitival Relatives can be interpreted with respect to the future:

(179) a. A man from Arkansas will be the first man to walk on Mars.
(≈ A man from Arkansas will be the first man who will walk on Mars.')

b. The first man to walk on Mars will be canonized in the year 3000.
(≈ The first man who will walk on Mars will be canonized in the year 3000.)
Plausible reading: *walk* < *canonization*
Implausible, but marginally available: *walk* = *canonization*
Unavailable: *canonization* < *walk*

c. The first man to walk on Mars will be American.
(≈ The first man who will walk on Mars will be American.)
Most plausible: *walk* = *american*
Plausible: *walk* < *american*
Unavailable: *american* < *walk*

Our basic claim is that the the future interpretation of the non-modal infinitival clause is not 'inherent' but is inherited from matrix clause. Therefore this interpretation is only available in the presence of a matrix future. In (180), there is no future in the matrix clause and it is not possible to interpret the infinitival clause shifted to the future. However, because of world knowledge we know that event of a man walking on Mars is not in the past/prsent. This is why (180) is judged as false.

(180) A man from Arkansas was the first man to walk on Mars.
≠ 'A man from Arkansas was the first man who will walk on Mars.'

Modal reading:
'A man from Arkansas was designated to be the first man who would walk on Mars.'

There is a future-oriented modal reading for (180), which can be paraphrased as *A man from Arkansas was supposed/designated to be the first man who would walk on Mars*. This future-oriented modal reading is distinct from the 'non-modal' future reading that we find in (179a-c). The future interpretations in (179a-c) do not assert that any particular person is supposed to walk on Mars. They just assert that whoever walks on Mars will have certain properties. The second difference lies in that the 'non-modal' future reading is either simultaneous or anterior to a future time determined by the matrix tense (cf. the unavailability of the *american* < *walk* readings in 179c). In the modal reading of (180), the modal future interpretation can plausibly be interpreted as picking out a time in the future i.e. the walking on Mars event could have been planned for 2030.

The future interpretation of the in (179a-c) follows from the availability of a matrix future to which the infinitival tense links up. In (179a), the *first* NP is in a predicative position and is therefore evaluated at a some point in the future. At that point of time, the Arkansan will be the first man to walk on Mars.

In (179b, c), the temporal variable of the subject DPs can be set independently of matrix tense. However, it has been noted before (cf. Enç 1986, and Musan 1995) that the temporal variable of a DP can be set to a time independent of matrix tense but that the option of independently setting the temporal variable to a future time is generally unavailable. So the temporal variable of the subject DP in (179b, c) can be independently set to a past time or to the present. The other option is that it is determined by the matrix future. The option of independently setting the temporal variable to a future point of time other than the matrix future is unavailable.

With the above set of independently motivated assumption, we can explain the distribution of 'non-modal' future interpretations in (181a-c).

(181) a. The first American to walk on the moon/#Mars visited my school yesterday.
(≈ The first American who walked on the moon visited ...)
(≠ The first American who will walk on Mars visited ...)

b. The first American to walk on the moon/#Mars is walking in my garden.

($\approx$ The first American who walked on the moon is walking in my garden.)
($\neq$ The first American who will walk on Mars is walking in my garden.)

c. The first American to walk on the moon/Mars will be canonized in 3000A.D.
($\approx$ The first American who walked on the moon will be ...)
($\approx$ The first American who will walk on Mars will be ...)

Due to its aspectual specification (perfective for non-statives) a Non-Modal Infinitival Relative clause can always be shifted into the past. This is why past readings are available in (181a-c) independent of the matrix tense specification. However, there is no corresponding 'covert' future in the aspectual specification of the infinitival clause and independent temporal variable setting a future point of time is generally unavailable. Therefore, for a non-modal future reading (cf. 181c), a matrix future is necessary.

7. Appendix A: Semantics of *only* and *first*

7.1. *only*

Semantics analyses of *only* differ in the number of arguments that *only* takes, what these arguments are, and how it takes these arguments. We survey two proposals here. In the first proposal, *only* takes an individual and a predicate as a complement. Both arguments are syntactically represented.

(182) a. Only Muriel voted for Hubert.
$[\forall x : x \neq Muriel]\neg vote(x, Hubert)$
$only(x,P) \Longleftrightarrow [\forall y : y \neq x]\neg P(y)$
Presupposition: $P(x)$ (Horn 1969)
$only(x,P)$ is true iff P is true of nothing other than x.

b. John only introduced $\mathbf{Bill}_F$ to Sue.
Bill, *only* move so that *Bill* is the first argument of only at LF. (Anderson 1969)

According to the second proposal, *only* takes a propositional first argument and a set of propositions as a second argument. The second argument is strictly speaking not represented syntactically.

(183) Let p be a proposition, C a set of propositions:
$only(p,C) \Longleftrightarrow p \wedge \forall q[q \neq p \wedge q \in C \Longrightarrow \neg q]$
presuppositional constraints on C: $p \in C \wedge \exists q \in C : q \neg = p$ (Rooth 1985, 1992)
$only(p,C)$ is true iff p is the only true proposition in C.

The value of C is constrained by focus in the following fashion:

(184) $only(C)(p\,\tilde{}\,C)$
where $p\,\tilde{}\,C$ adds a presupposition that C is a subset of the focus value of p, $\tilde{}\,C$ has no other effect on meaning, $p\,\tilde{}\,C$ denotes the same object as p.

(185) provides an illustration of this system.

(185) John only introduced $\mathbf{Bill}_F$ to Sue. ($= p$)
FocusValue(p) = {John introduced Tom to Sue, John introduced Teresa to Sue, John introduced Bill to Sue, John introduced Sarla to Sue, ..., }
Assume C = FocusValue(p), then
$only(C)(p) \Longleftrightarrow$ "John introduced Bill to Sue" is the only true proposition in C.

7.2. Ordinals ('first', 'second', ..., 'last')

The semantics proposed in sect. 6.1 is a two argument semantics, repeated here as (186).

(186) $first(P) = \lambda x \lambda t.[P(x,t) \wedge \forall y[[x \neq y \wedge \exists t'.P(y,t')] \Longrightarrow F(P)(x) < F(P)(y)]]$
where
$F(P)(x) = \iota \lambda t.[P(x,t) \wedge \forall t' < t.\neg P(x,t')]$

If we avail ourselves of a contextually restricted set of properties of times C, we can define a one argument analysis of *first*:

(187) $first(C)(P) = \lambda t.[P(t) \wedge \forall P'[[P' \neq P \wedge P' \in C] \Longrightarrow F(P) < F(P')]]$
where
$F(P) = \iota \lambda t.[P(t) \wedge \forall t' < t.\neg P(t')]$

8. Appendix B: *in situ* licensing of non-modal infinitives

The following two steps are required in order for the licensing configuration as sketched out in the main text to obtain: (i) the head NP of a non-modal infinitival relative has to be interpreted internal to the relative clause, and (ii) the licensor moves out of the head NP which is in subject position to a position from where it takes the infinitival clause as its complement and can license the non-modal reading. We will now consider an alternative approach under which the *first/-est/only NP* is not required to reconstruct into the relative clause for licensing purposes. The licensing configuration is essentially met on the surface structure. This is shown in (188),

(188) (structures indicated are intended to be both surface structures and LFs)

a. the [first [$_{NP}$ [$_{NP}$ man] [$_{RC}$ to walk on the moon]]]

b. the [only [$_{NP}$ [$_{NP}$ man] [$_{RC}$ to walk on the moon]]]

c. the [tallest [$_{NP}$ [$_{NP}$ man] [$_{RC}$ to walk on the moon]]]

For a non-modal reading to be licensed, the licensor must take the infinitival clause as a complement.

The head NP *man* may be interpreted internal to the relative clause; what distinguishes this alternative proposal is that the licensor (*first/-est/only*) stays *in situ*, external to the relative clause for the purposes of licensing the non-modal interpretation.

Let us now go over the properties of non-modal infinitival relatives that the analysis in the main text was able to explain and see how well the *in situ* licensor analysis does at capturing them.

The first property concerns the fact that only superlatives, ordinals and nominal *only* and not numeral quantifiers, *few/many* etc. are able to license non-modal readings. The explanation given under the reconstruction analysis (in the main text) was as follows: *Few/many*/numeral quantifiers are unable to appear in the licensing configuration since their semantics does not involve them taking the infinitival clause as a complement. Hence they are unable to license the non-modal interpretation. We can transport this explanation essentially unchanged to the *in situ* analysis. Numerals/*few/many* are unable to license the non-modal interpretation from their *in situ* position in *#the two/few/many men to walk on the moon* because of their semantics. Numerals/*few/many* do not take the infinitival clause as a complement; they combine with the predicate denoted by the [NP CP] via Predicate Modification (set intersection). So numerals/*few*/*many* are unable to license the non-modal interpretation under the *in situ* analysis for the same reasons as under the Reconstruction analysis.

The fact that the licensor of the non-modal reading was unable to associate with focus when it was involved in licensing a non-modal reading was predicted by the Reconstruction analysis. It turns out that we can provide a simple explanation for this fact under the *in situ* analysis also. Consider the LFs in (189).

(189) **Joan**$_F$ gave Mary the most expensive telescope to be built in the 9th century.

a. LF with Association with focus
$-est$ λd [**Joan**$_F$ gave Mary the d-expensive telescope to be built in the 9th century].

b. LF with no association with focus
Joan$_F$ gave Mary the [most [[expensive present] [to be built in the 9th century]]]

(189a) represents the case where the superlative licensor associates with matrix focus. But then it is in the wrong configuration to license the non-modal reading of the infinitival clause. For the non-modal reading to be available the infinitival clause needs to be the complement of the licensor. Therefore the LF in (189a) is ruled out and only the LF in (189b) is available.

The reconstruction analysis is able to provide an explanation for why both relative clause-external nominal ordinals and relative clause-internal adverbial ordinals were both able to license a non-modal interpretation. The *in situ* analysis is unable to capture these two kinds of licensing in a uniform manner. This may, however, not be such a bad thing because for many speakers the licensing of the non-modal reading from inside the relative clause does not seem to be as robust as licensing from outside the relative clause. If we set aside the cases of licensing by relative clause-internal ordinals aside, the *in situ* analysis is able to provide a very simple and intuitive explanation for why relative clause-internals superlatives and nominal *only* are unable to license a non-modal reading.

(190) a. # The student to scream the loudest

b. # The boy to give Mary the only telescope

These cases are bad under the *in situ* analysis because the superlative/*only* is not in the right position either in overt syntax or at LF. Even after LF-movement, the superlative/*only* would not reach the position occupied by the licensor when it occurs relative clause-externally.

The reconstruction analysis is also able to explain why the class of licensors of the non-modal interpretation did not include any quantifiers cf. * *every man to walk on the moon*. The explanation hinged on the fact that quantifiers were generated in D^0 and did not reconstruct into the relative clause. Thus they were never in a local enough configuration with the infinitival clause to license it. The *in situ* analysis cannot use this difference between quantifiers and the class of non-modal infinitival relative licensors. The relevant difference has to be found somewhere else, perhaps in the semantics of superlatives, ordinals, and nominal *only*.

To conclude: giving up on the assimilation of the licensing of a non-modal interpretation from a relative clause-internal and from a relative clause-external position open the door to an *in situ* analysis. The empirical consequences of the *in situ* analysis and the reconstruction analysis overlap to a considerable degree. On the side of the reconstruction analysis is the fact that it is able to explain (i) the data regarding the licensing of the non-modal reading by a relative-clause internal ordinal, and (ii) the fact that the class of licensors is a subclass of elements that can reconstruct. The *in-situ* analysis is, on the other hand, able to provide a simple explanation for why relative clause-internal superlatives and *only* are unable to license non-modal interpretations.

9. Conclusions

We have shown that the non-modal reading of infinitival relative clauses requires a licensor. The class of licensors includes superlatives, ordinals, adnominal *only*. We proposed that the licensing configuration for Non-Modal Infinitival Relatives obtains under reconstruction into the relative clause. The fact that focus-sensitive elements lose their focus sensitivity when involved in licensing a non-modal reading is shown to follow from the licensing proposal.

The temporal semantics of *first* and its interaction with non-modal relative clauses is investigated. To enable this investigation, a semantics for *first* is provided. It is shown that *first* takes as an argument a property of times and individuals. The result of applying *first* is another property of times and individuals.

We identified a constraint on the temporal interpretation of the head NP - in aspectually unspecified infinitivals, the head NP has to be true at the time of event denoted by the infinitival clause. This constraint on temporal interpretation was also shown to follow from the licensing configuration, which requires reconstruction of the head NP into a relative clause internal position. Further examination of the temporal properties of overtly aspectually unspecified infinitival clauses prompted a recognition of their aspectual properties. In particular, unspecified non-stative predicates are treated as having perfective aspect. Various aspects of their temporal interpretation and interaction with the semantics of *first* are shown to follow from this proposal.

Chapter 4
Infinitival questions

Where do infinitival questions occur? Are there predicates that exclusively subcategorize for infinitival questions? What are the properties of the modality that appears in infinitival questions? These are some of the questions addressed in this chapter. The distribution of infinitival questions is examined in Sect. 1. In Chapter 2, we observed that infinitival questions always involve modality. The infinitival [+wh] complementizer was identified as the source of the modality. We will now explicate the properties of the modality that appears in infinitival questions. It will be shown that the modality in infinitival questions is deontic/bouletic except in a limited set of environments where it can be circumstantial. This modality can never be epistemic (Sect. 2.1). We note that the force of the modality in an infinitival question seems to vary. In some cases, a paraphrase with an existential modal (e.g. *could*) seems appropriate and in others, a paraphrase with a universal modal (e.g. *should*) seems appropriate. Sect. 2.2 discusses the factors that seem to determine the force of the infinitival modality.

Sect. 3 provides an explanation for the variability in the modal force of the infinitival modality. The basic proposal is that there is a single infinitival modality. This modality is neither an existential modality nor a universal modality. A detailed discussion is provided of how various factors make this infinitival modality equivalent in certain environments to a deontic *should*, in others to a circumstantial *could* etc. The proposal developed for infinitival questions is applied in Sect. 4 to handle variable force effects in infinitival relative clauses.

1. Infinitival questions: distribution and subcategorization

We will begin with a discussion of the distribution of infinitival question complements. Not all predicates that take finite question complements take infinitival question complements. However, all predicates that take infinitival question complements take finite question complements. The distribution of infinitival question complements (Sect. 1.1.2) will be compared to the distribution of finite question complements (Sect. 1.1.1).

Some but not all of the predicates that take infinitival question complements take infinitival non-question complements. We will examine whether the cases of predicates that take both infinitival question complements and infinitival non-question complements involve two distinct predicates - one that takes an infinitival non-question complement and one that takes an infinitival question complement, or whether we have one predicate that takes both infinitival question and non-question complements. This question will be explored in the general context of subcategorization and the generalizations that hold regarding predicates that take infinitival questions (Sect. 1.4).

1.1. The distribution of infinitival questions

Before we start looking at predicates that take infinitival question complements, let us briefly go over a classification of predicates that take finite question complements.

1.1.1. Predicates that take finite question complements

The following list provides a classification of the different kinds of predicates that take finite interrogative complements. It builds heavily upon a classification provided in Karttunen (1977, pg.6) and is not meant to be exhaustive.

(191) a. Verbs of retaining knowledge:
know, be aware, recall, remember, forget

b. Verbs of acquiring knowledge:
learn, notice, find out, discover

c. Decision verbs:
decide, decide on, determine, specify, agree on, control

d. Verbs of conjecture:
guess, predict, bet on, estimate

e. Opinion verbs:
agree about, be certain (about), have an idea (about), be convinced about

f. Verbs of relevance:
matter, be relevant, be important, care, be significant

g. Verbs of dependency:
depend on, be related to, have an influence on, be a function of, make a difference to

h. Verbs of one-way communication:
tell, show, indicate, inform, disclose

i. Verbs of two-way communication:
discuss, squabble over, talk about

j. Verbs of cogitation:
address (the issue of), consider, debate, deliberate, fret about, study

k. Inquisitive verbs:
ask, wonder, investigate, be interested in

In addition, several nouns also take finite interrogative complements e.g. *question of, disagreement over/about, ideas about, accounts of, inquiries about, controversy over, discussion of* etc.

1.1.2. Infinitival questions

Unlike finite question, infinitival questions (with a few exceptions)[49] seem to be restricted to embedded environments. We begin with some naturally occurring examples of infinitival interrogative complements extracted from the Wall Street Journal corpus. These examples illustrate the classification of question-embedding predicates in (191).

(192) a. Verbs of retaining knowledge: *forget, know*
Mr. Islas knows whom to thank for it all.
He knows how to steal computer secrets. He knows how to embezzle money with the help of computers. He knows how to plant viruses 'in computer programs and thereby create data-processing havoc for their users.
It allegedly makes hair dryers, but nobody knows where to buy one.
Burger King's nervous franchisees, still upset over the disastrous Herb the Nerd promotion that bombed two years ago, are wondering whether the chain also has forgotten how to pitch hamburgers.

b. Verbs of acquiring knowledge: *discover, figure out, learn*
   Mr. Rothchild took his precious savings and spent a year trying to discover where to invest them.
c. Decision verbs: *choose, decide, rule*
   As it stands now, airlines decide when to retire their planes.
   Twenty percent of the electorate decided who to vote for in the last week of the campaign.
d. Verbs of conjecture: *No examples in the corpus*
e. Opinion verbs: *be baffled about, be divided over, be (un)sure*
   Western companies weren't sure who to talk to.
   Prentice-Hall Inc. is unsure whether to continue using 5 1/4-inch disks or the newer 3 1/2-inch ones.
f. Verbs of relevance: *No examples in the corpus*
g. Verbs of dependency: *No examples in the corpus*
h. Verbs of one-way communication: *advise on, dictate, explain, give advice on, say, show, tell*
   With the first generation of personal computer programs, users told computers what to do and when to do it. Now there 's software that enables computers to tell people what to do and when to do it.
   In an argument it 's important to know where people are coming from before you tell them where to go.
i. Verbs of two-way communication: *argue over, debate, discuss, talk about, squabble over*
   Mr. Kurnit notes that studios and cable operators continue to squabble over how to run the business and divide pay-per-view revenues.
   Pentagon officials are discussing whether to extend protection to all neutral ships.
j. Verbs of cogitation: *address, consider, deliberate, study*
   Tom Gildersleeve, the Air Force 's top civilian expert on electricity supply, says he is considering whether to seek out independent power producers.
   They address how to organize a world-wide company with many different profit centers so that it brings all its resources to bear on the needs of each customer.

Under the plan, the EPA would study where to create markets for recycled products.

k. Inquisitive verbs: *ask, wonder*
He wondered how to make a living from grapes, and had a brainstorm: champagne.
Slip a song into a classical broadcast and listeners by the dozen call to ask where to buy the album.

There were no examples in the corpus of infinitival question complements of verbs of conjecture (*guess, predict* etc.), verbs of relevance (*be relevant, be important* etc.), and verbs of dependency (*depend on, have an influence on*). This absence is not an accident of the corpus. Infinitival question complements of the predicates in question are either totally ungrammatical or at least quite marginal.

(193) a. verbs of conjecture
  i. * John predicted who to invite to the party.
  ii. ??? John guessed who to talk to at the party.
b. verbs of relevance
  i. * It matters who to invite to the party.
  ii. * It is important who to talk to at the party.
c. verbs of dependency (exs. from Lahiri 1998, pg. 51)[50]
  i. ?? What to do depends upon where to be.
  ii. ?? Who comes to Boston depends upon how to get there from LA.

Emotive predicates like *be surprising*, *be amazing* etc. also do not take infinitival question complements.

(194) a. * It is amazing what to do. (vs. It is amazing who Bill knows)
b. * It is surprising what to do. (vs. It is surprising how much Bill has accomplished)

Elliott (1971) and Grimshaw (1977) do not consider the complements of these verbs to be true interrogatives. They argue that the complements of these verbs form a distinct class which they call 'exclamatives'. From their perspective, the fact that emotive predicates do not take infinitival question complements follows from the fact that infinitival *wh*-clauses do not form good 'exclamatives'. They do not provide an explanation for this last fact.

We will return to the question of why infinitival questions are degraded or unacceptable with the aforementioned classes of predicates. Note that while there seem to be predicates that take finite questions but not infinitival questions, there are no predicates that take infinitival questions but not finite questions.

A fairly large number of nouns take infinitival question complements e.g. *accounts of, advice on, agreement about, befuddlement about, challenge of, choice of, conception of, consensus about, controversy over, decision on, details of, differences on, disagreement over/about, discussion of, emphasis on, ideas about, information on, inquiries about, instructions on, issue of, message about/on, plan for, question of, seminars on, specifics of, standards on, stories about, study of, talk about, tactics of, tiff over, tradition of, view of* etc.

(195) a. At the very least, the change could deprive investors of a choice of how to pay for mutual funds.

b. NATO defense ministers postponed a decision on how to share the $500 million burden of moving a U.S. fighter wing from Spain to Italy.

c. Pick up any newspaper travel section and you'll find accounts of swarms of tourists at the South Pole, overcrowding in the Galapagos, overcharging in Lhasa, where to eat at midnight in Slovenia.

d. This time, Chinese economic planners can't reach a consensus about how to tackle inflation.

e. Miller is also struggling with the critical issue of how to put Miller High Life back on track, a challenge that thus far has eluded both ad agencies and company executives.

1.2. Infinitival questions and *wh*-phrases

In our discussion of infinitival relatives, we noted an unexplained restriction on the material that could appear in the [Spec,CP] of such relative clauses: overt material is only possible if embedded in a PP.

(196) Infinitival Relative Restriction: only PPs may appear in [Spec,CP].

a. a knife [Op_i [PRO to cut the bread with t_i]]

b. * a knife [$which_i$ [PRO to cut the bread with t_i]]

c. a knife [[with which]$_i$ [PRO to cut the bread t$_i$]]

This restriction does not apply to infinitival questions. In fact, the cases with pied-piping feel somewhat stilted.

(197) a. Jan knows [which knife to cut the bread with].

b. ?Jan knows [with which knife to cut the bread].

Given that the null operator option is unavailable with constituent questions, infinitival or finite, the fact that (196) does not apply to infinitival questions makes a certain amount of sense. If it did, most infinitival questions would be ineffable. We would only be able to construct infinitival questions that involved abstraction over PP positions.

But infinitival questions have a mysterious restriction of their own. One cannot construct infinitival questions with the interrogative pronoun *why*. This is a restriction specific to infinitival questions. Finite *why* questions are fine.

(198) a. *Michael wonders why to do that.
(vs. Michael wonders why he should do that.)

b. *I don't know why to make such a fuss.
(vs. I don't know why I should make such a fuss.)

It might seem like this is a more general restriction involving infinitival abstraction. The relative/free relative cases corresponding to (198) are also ungrammatical.

(199) a. Infinitival Relative: *[the reason [why to do that]] remains a mystery.

b. Infinitival Free Relative: *[Why to do that] remains a mystery.

But the ungrammaticality of (198a, b) is not due to the presence of *why*. (199a) is ungrammatical due to the (196) and (199b) is ungrammatical because infinitival free relatives seem to be systematically unavailable in English, irrespective of the *wh*-expression.

We do not have any explanation of why *why*-infinitival questions are impossible in English. Initial explorations indicate that they are available in Spanish. But we note that within English and also crosslinguistically *why* often does not pattern with other *wh*-words. In English, matrix *wh*-infinitivals are marked but to the extent they are possible, they need an infinitival *to*. In contrast, matrix infinitival *why* questions are quite common but these always lack the infinitival *to*.

(200) a. What to do?/*What do?

b. Why worry?/*Why to worry?

In the Indo-Aryan language Hindi-Urdu, reduplication of *wh*-pronouns is a productive process. But it does not apply to *kyõ* 'why'. Other *why*-specific restrictions are noted in Ko (2005), who tries to derive some of these restrictions from base-generation *why* in the C-domain pursuing a suggestion made for French *pourquoi* on page 47 of Rizzi (1990).

1.3. The crosslinguistic distribution of infinitival questions

In the previous sections, we have examined the distribution and form of infinitival questions in English. Such an exploration presupposes that English has infinitival questions. Curiously there are languages that lack infinitival questions.

Somewhat trivially this is the case for languages without an infinitive such as Greek. But this is also the case for Hindi-Urdu, which does have a category commonly analyzed as an infinitive. As noted in Butt (1995) and Dayal (1996), infinitival clauses in Hindi-Urdu do not form a scope domain. While *wh*-phrases in finite clauses can take scope within the finite clause, this is never an option with infinitival clauses.

(201) a. *wh*-in-situ in finite clause: only embedded scope

Mona-ne faislaa kar liyaa hai [ki vo
Mona-Erg decision.m do TAKE.Pfv.MSg be.Prs.Sg that she
kis-se mil-egii].
who-with meet-Fut.FSg

'Mona has decided who she will meet.'

b. overt *wh*-movement out of finite clause: matrix scope strongly favored

kis-se$_i$ *Mona-ne faislaa kar liyaa hai*
who-with Mona-Erg decision.m do TAKE.Pfv.MSg be.Prs.Sg
[ki vo t_i *mil-egii]?*
that she meet-Fut.FSg

'Who has Mona decided that she will meet?'
(embedded scope is also available, but not favored)

c. infinitival clause: only wide-scope: only matrix scope

Mona-ne kis-se mil-ne-kaa faislaa kar
Mona-Erg who-with meet-Inf.Obl-Gen decision.m do
liyaa hai?
TAKE.Pfv.MSg be.Prs.Sg

'Who has Mona decided to meet?'

Along with infinitival questions, Hindi also seems to lack any true infinitival relatives. Thus there are no object infinitival relatives but there are subject infinitival relatives, which are plausibly analyzed as reduced relatives. Like other reduced relatives in Hindi-Urdu, the modifiers in question are prenominal and lack any overt relative pronouns. Additionally they require the nominalizer -*vaalaa*.

(202) a. no object relatives:

**[Varun-kaa ___ kaaṭ-ne]-(vaalaa) peṛ*
Varun-Gen cut-Inf.Obl-NMLZ tree

intended: 'a/the tree for Varun to cut'

b. subject relatives, presumably reduced:

[___ peṛ kaaṭ-ne]-vaalaa laṛkaa
tree cut-Inf.Obl-NMLZ boy

'The boy who cuts trees'

There is thus no evidence in Hindi-Urdu that there is ever any instance of A'-movement that terminates in an infinitival clause. We have already seen that there are no embedded infinitival questions or infinitival relatives. We note that there are also no matrix *wh*-infinitivals either. These facts make sense in the light of Butt (1995)'s observation concerning the nominal nature of Hindi-Urdu infinitives. She analyzes them as gerunds and one aspect of gerunds is that they lack a C-domain. If question formation and full relative clause formation require a C-domain, the absence of infinitival questions and infinitival relatives in Hindi-Urdu, reanalyzed as gerunds , follows.

A more curious case is that of German, which like Hindi-Urdu lacks embedded infinitival questions and infinitival relative clauses but which, as Reis

(2003) has shown, has matrix *wh*-infinitivals. The German facts have an additional twist introduced by the infinitival marker *zu*. In embedded environments, *wh*-infinitives are systematically ill-formed. Matrix environments permit *wh*-infinitivals but these infinitives cannot have *zu*.

(203) German (from Reis (2003))

a. embedded infinitival question

Ich weiß nicht, was zu kaufen.
I know not what to buy
'I don't know what to buy.'

b. matrix infinitival question with *zu*

**Wohin zu gehen?*
where to go
'where to go'

c. matrix infinitival question without *zu*

Wohin gehen?
where go
'Where should/could/may we/I/one go?'

Reis (2003)'s proposal for German has *zu* being a marker of subordination that is incompatible with the presence of a [+wh] feature on C^0. If the *wh*-infinitival is not embedded, *zu* is not permitted and hence does not block the possibility of a [+wh] C^0.

To sum up, languages vary in whether they permit infinitival questions and infinitival relatives and if they do so, where. We can think of this variation in terms of variation in the properties of the C-domain associated with infinitival clauses in various languages. English is the most flexible and freely allows for a [+wh] C^0 to be associated with an infinitival clause. At the other extreme we have Hindi-Urdu, which disallows [+wh] C^0's with infinitival clauses. German falls somewhere in between. Infinitival clauses there may have a [+wh] C^0 but only in the absence of the infinitival marker *zu*.

1.4. Infinitival questions and subcategorization

We will now consider the subcategorizational properties of the predicates that take infinitival question complements. The goal is to find the generalizations that exist regarding the kind of complements the predicates that take infinitival question complements take and to explain such generalizations.

1.4.1. Non-interrogative infinitival complements

It is not the case that all predicates that can take infinitival question complements can take infinitival non-question complements. e.g. *know* takes infinitival question complements but not non-interrogative infinitival complements. Here we will divide the predicates that take infinitival question complements into two groups on the basis of whether they take non-interrogative infinitival complements.

(204) a. take non-interrogative infinitival complements
verbs of retaining knowledge: *forget, remember*
verbs of acquiring knowledge: *learn*
verbs of decision: *choose, decide*
verbs of one-way communication: *advise, instruct, tell*
inquisitive verbs: *ask**

b. do not take non-interrogative infinitival complements
verbs of retaining knowledge: *know*
verbs of acquiring knowledge: *calculate, determine, establish, figure out*
verbs of cogitation: *deliberate on, focus on, puzzle over, study*
verbs of one-way communication: *dictate, explain, show*
verbs of two-way communication: *discuss, squabble over, talk about*
inquisitive verbs: *wonder*

1.4.2. Some subcategorizational generalizations

The question of whether a particular predicate takes both finite and infinitival question complements is part of the more general question of what an arbitrary predicate can subcategorize for. If we restrict ourself to finite and infinitival clausal complements which may or may not be questions, we find the distribution in (205). The feature clusters are meant to be interpreted as exhaustive - so `[-wh,+inf]` in (205k) represents the class of predicates that exclusively subcategorizes for non-interrogative infinitival clauses.

(205) a. `[-wh,-inf] [-wh,+inf] [+wh,-inf] [+wh,+inf]`: *forget, decide, tell*
b. `[-wh,-inf] [-wh,+inf] [+wh,-inf]`: *be important*
c. `[-wh,-inf] [-wh,+inf] [+wh,+inf]`: not attested
d. `[-wh,-inf] [+wh,-inf] [+wh,+inf]`: *know*
e. `[-wh,+inf] [+wh,-inf] [+wh,+inf]`: *ask**
f. `[-wh,-inf] [-wh,+inf]`: *convince, persuade*
g. `[-wh,-inf] [+wh,+inf]`: not attested
h. `[-wh,+inf] [+wh,+inf]`: not attested
i. `[-wh,+inf] [+wh,-inf]`: ?
j. `[+wh,-inf] [+wh,+inf]`: *wonder*
k. `[-wh,-inf]`: *believe, deny*
l. `[-wh,+inf]`: *try*
m. `[+wh,-inf]`: *(?)investigate*
n. `[+wh,+inf]`: not attested

There are many gaps in the classification in (205). We can group the classes in (205) into three subgroups: predicates that take infinitival questions, predicates that do not take infinitival questions, and unattested subcategorization patterns.

(206) a. predicates that take infinitival questions:
i. `[+wh,-inf] [+wh,+inf]`: *wonder*
ii. `[-wh,-inf] [+wh,-inf] [+wh,+inf]`: *know*
iii. `[-wh,+inf] [+wh,-inf] [+wh,+inf]`: (*ask**, *choose*)
iv. `[-wh,-inf] [-wh,+inf] [+wh,-inf] [+wh,+inf]`: *forget, decide, tell*
b. predicates that do not take infinitival questions:
i. `[-wh,-inf] [-wh,+inf] [+wh,-inf]`: *be important*
ii. `[-wh,-inf] [-wh,+inf]`: *convince*
iii. `[-wh,-inf]`: *believe, deny*
iv. `[-wh,+inf]`: *try*

v. `[+wh,-inf]`: *investigate*

c. unattested subcategorization patterns:

i. `[-wh,-inf] [-wh,+inf] [+wh,+inf]`

ii. `[-wh,-inf] [+wh,+inf]`

iii. `[-wh,+inf] [+wh,+inf]`

iv. `[+wh,+inf]`

v. `[-wh,+inf] [+wh,-inf]` (??)

The following generalization emerges: if a predicate takes infinitival questions, it also take finite questions. This generalization, which we will call *The No Subcategorization for Infinitival Questions* generalization, is somewhat unexpected, in particular, since there are predicates that take infinitival non-question complements but not finite non-question complements e.g. *try*. It follows if we assume that there is s-selection, but no c-selection, and that infinitival questions and finite questions denote the same kind of semantic object. Then if a predicate s-selects for the kind of semantic object an infinitival question denotes, it also s-selects for finite questions. [51]

We now have to explain why all predicates that take finite question complements do not also take infinitival question complements. We propose that this asymmetry (take infinitival question → take finite question, take finite question ↛ take infinitival question) exists because of a difference between the interpretational possibilities available to finite questions and infinitival questions. In contrast to finite questions which can denote sets of any kind of propositions, infinitival questions are restricted to denoting sets of a restricted class of modal propositions: propositions involving the covert modality contributed by the infinitival [+wh] complementizer. A predicate that take infinitival questions subcategorizes for a particular kind of question meaning. This meaning can be rendered by some finite modal question or the other. The meanings denoted by all finite modal questions cannot be rendered by infinitival questions. Consequently, it is possible for a predicate to take a finite question and not take an infinitival question but not vice-versa. This is, of course, at this point only a suggestion. A proper account would need to isolate the meanings that cannot be captured by an infinitival questions and show why the semantics of certain verbs prevents them from combining with infinitival question. This task is left for future research.

1.4.3. One predicate or many

The following question arises when we look at a classification like (206): When there is a predicate that subcategorizes for `[-wh,-inf]` `[-wh,+inf]` `[+wh,-inf]` `[+wh,+inf]`, do we really have one predicate that subcategorizes for `[-wh,-inf]` `[-wh,+inf]` `[+wh,-inf]` `[+wh,+inf]` or do we have four (or possibly more) distinct predicates that subcategorize for subsets of { `[-wh,-inf]` `[-wh,+inf]` `[+wh,-inf]` `[+wh,+inf]` }?

It is not easy to answer this question in the general case. The default strategy seems to be to assume that there is one predicate that subcategorizes for all the different kinds of clausal complements. If this strategy does not work e.g. because it is not feasible to assign a common meaning to the predicate in its different incarnations, we postulate the existence of two (or more) distinct predicates.

We will consider the three classes of predicates in (206a) to see whether we can provide a unified semantics to the predicates that works irrespective of what kind of complement they take. But before we do that let us review a line of work that provides a semantics for predicates like *know* that works whether they take a finite indicative complement or a finite interrogative complements.

(207) a. Herb knew [that Hafdis liked Ásta].

b. Herb knew [who Hafdis liked].

The proposal is that predicates like *know* semantically subcategorize for a proposition. Interrogative complements of *know* have the same type as propositions (cf. Groenendijk & Stokhof (1982, 1984)), or they are interpreted as open propositions (cf. Berman (1991)), or a rule of Interrogative-Raising is proposed, which leaves behind a propositional trace (cf. Lahiri (1991, 1998)). The net result is that at the relevant interpretive level, the complement of *know* is a proposition irrespective of whether in overt syntax, *know* takes a interrogative clause as a complement or a declarative clause.

From our explanation of the *No Subcategorization for Infinitival Questions* Generalization, we have it that infinitival and finite question complements must denote the same kind of object. Hence the predicates that subcategorize for `[+wh,-inf]`, `[+wh,+inf]` e.g. *wonder* can be provided a unified semantics. Since it does not seem to be the case that *wonder* has a special meaning when it take infinitival question complements, we can take whatever semantics we have for $wonder_{+wh,-inf}$ and use it for $wonder_{+wh,+inf}$.

Next, we come to the predicates that subcategorize for `[-wh,-inf]`, `[+wh,-inf]`, `[+wh,+inf]` e.g. *know*. We have seen that there are several proposals which give $know_{-wh,-inf}$ and $know_{+wh,-inf}$ a common semantics. By our assumption that infinitival questions and finite questions denote the same kind of semantic object, we can therefore provide all the predicates in the class in (206a.ii) a semantics that works whether they take an indicative clause, a finite interrogative clause, or a non-finite interrogative clause as a complement.

Finally, we come to the classes in (206a.iii, iv). Both these classes involve `[-wh,+inf]` i.e. non-interrogative infinitival complements. We have seen that assimilation of $PR_{-wh,-inf}$ and $PR_{+wh,-inf}$ is feasible. The assimilation of $PR_{+wh,-inf}$ and $PR_{+wh,+inf}$ is also feasible. The question then is the following: can we give a semantics to PR that will work for `[-wh,+inf]` in addition to the others?

We will go through the list of predicates in (204a) and examine the relationship between $PR_{-wh,+inf}$ and $PR_{+wh,+inf}$. Since it was feasible to provide a common semantics for $PR_{-wh,-inf}$ (finite indicative clauses) and $PR_{+wh,-inf}$ (finite interrogative clauses), on an analogy, we will check if it is possible to provide a common semantics for $PR_{-wh,+inf}$ and $PR_{+wh,+inf}$.[52] In order to retain our explanation for the *No Subcategorization for Infinitival Questions* Generalization, a common semantics for $PR_{-wh,+inf}$ and $PR_{+wh,+inf}$ should extend to $PR_{+wh,-inf}$ (finite interrogatives) also.

The first set of predicates that we examine are predicates like *forget* and *remember*. $forget_{-wh,+inf}$ and $remember_{-wh,+inf}$ are negative and positive implicative verbs respectively (cf. Karttunen 1971).

(208) a. Olafur forgot to lock his door.
$\rightarrow$ Olafur didn't lock his door.

b. Olafur remembered to lock his door.
$\rightarrow$ Olafur locked his door.

$forget_{+wh,+inf}$ and $remember_{+wh,+inf}$ are not implicative verbs. In this respect, they are similar to $forget_{+wh,-inf}$/$forget_{-wh,-inf}$ and $remember_{+wh,-inf}$/$remember_{-wh,-inf}$ respectively.

(209) a. Olafur forgot which door to lock.
Olafur forgot which door he was supposed to lock.
Olafur forgot that he was supposed to lock the Blue Room.
$\not\rightarrow$ Olafur didn't lock the door.

b. Olafur remembered which door to lock.
Olafur remembered which door he was supposed to lock.
Olafur remembered that he was supposed to lock the Blue Room.
$\not\rightarrow$ Olafur locked the door.

The semantics of $forget_{-wh,+inf}$ and $remember_{-wh,+inf}$ seems distinct enough from $forget_{+wh,+inf}$ and $remember_{+wh,+inf}$ to preclude giving them identical lexical entries right now. On the other hand, it seems reasonable to assimilate $forget_{+wh,+inf}$ and $remember_{+wh,+inf}$ to the *forget* and *remember* that take finite complements.

Next we turn to predicates like *ask*. It is very likely that $ask_{-wh,+inf}$ is only accidentally homophonous with $ask_{+wh,+inf}$.

(210) a. Stefan Arni asked Hafdis$_i$ [PRO$_i$ to leave].

b. Stefan Arni$_i$ asked Hafdis [when PRO$_i$ to leave].

$Ask_{+wh,-inf}$ is object control while the $ask_{+wh,+inf}$ is subject control. $Ask_{+wh,-inf}$ and $ask_{+wh,+inf}$ do not seem to have a unified meaning - $ask_{+wh,-inf}$ means something like *request*, while the second $ask_{+wh,+inf}$ means something like *put a question to*.

The last set of predicates we examine includes *choose, learn, decide, tell*. In contrast to *ask* and *forget/remember*, the other verbs that take both non-interrogative and interrogative infinitival complements seem more amenable to a unified treatment of $PR_{-wh,+inf}$ and $PR_{+wh,+inf}$.[53]

(211) a. verbs of acquiring knowledge

i. Stefan Arni$_i$ learned [PRO$_i$ to blend blue paint with black pigment] in Siena.

ii. Stefan Arni$_i$ learned [what paint PRO$_i$ to blend with black pigment] in Siena.

b. verbs of decision

i. Stefan Arni$_i$ has decided [PRO$_i$ to leave at noon].

ii. Stefan Arni$_i$ has decided [when PRO$_i$ to leave].

c. verbs of one-way communication

i. Stefan Arni advised/instructed/told Hafdis$_i$ [PRO$_i$ to leave at noon].

ii. Stefan Arni advised/instructed/told Hafdis$_i$ [when PRO$_i$ to leave].

An explicit semantics for *choose, learn, decide, tell* that works for all their different kinds of complements will not be provided here. However, there do not seem to be any obvious obstacles in the way of providing such an explicit semantics.

The goal of this section was to examine, for the predicates that took both infinitival question complements and infinitival non-question complements, whether these two usages, call them $PR_{-wh,+inf}$ and $PR_{+wh,+inf}$ could be given a unified treatment. The conclusion is that the answer is sometimes no (*forget, remember, ask*) and sometimes yes (*choose, learn, decide, tell.* In general for all predicates PR, for which $PR_{-wh,-inf}$ and $PR_{+wh,-inf}$ exist, we find that a common semantics for these two uses is available. This is not case for all predicates PR, for which $PR_{-wh,+inf}$ and $PR_{+wh,+inf}$ exist.

2. Modality

All infinitival questions involve modality. Question formation involves A'-movement to the specifier of a [+wh] complementizer. In the case of infinitival questions, the A'-movement is to the specifier position of an infinitival [+wh] complementizer. This complementizer has modal semantics; it is the source of the modality in an infinitival question. More generally, it is due to the presence of such a complementizer that every structure that has A'-movement that terminates in an infinitival clause involves modality (cf. Chapter 2).

2.1. The nature of infinitival modality

Given that infinitival questions are always modal, we have to address the question of what the nature of this modality is. Modals come in many flavors - there is an issue of the kind of modality we are dealing with e.g. ability, bouletic, circumstantial, deontic, epistemic etc. and the force of the modality - existential, universal, or somewhere in between. In this section, I will address the question of the kind of modality that we find in infinitival questions setting aside the question of the force of infinitival modality for the next section.

The modality in an infinitival question is, for the most part, deontic/bouletic i.e. in consonance with certain laws, conventions, or desires. In certain well-defined cases, it can be circumstantial modality. It is, however, never epistemic modality.

Consider the range of interpretations available to finite modal and finite modal question complements to *know*. In order to bring out the epistemic reading, stative predicates will be used. We will see that depending upon the aspectual properties of the modal clause one reading or other is harder to get.[54]

(212) declarative complement

a. John knows that he must be at home at 9pm.
Reading 1: *Deontic*: It is necessary for the satisfaction of goals/by the law that John be at home at 9pm. John knows this.
Reading 2: ?? *Epistemic*: John is an amnesiac. He is trying to put together his past life by looking at various pieces of evidence. From the evidence it follows that he is at home every day at 9pm. John knows this.

b. John knows that he must have been at home at 9pm.
Reading1: ?? *Deontic*: It is necessary for the satisfaction of the goals/by the law for John to have been at home at 9pm. John knows this.
Reading 2: *Epistemic*: John is an amnesiac. From the evidence it follows that he was at home at 9pm. John knows this.

c. John knows that he could/may be at home at 9pm.
Reading 1: *Deontic*: It is compatible for the satisfaction of the goals/with the law that John be at home at 9pm. John knows this.
Reading 2: ?? *Epistemic*: It is compatible with the facts/ the way the world is that John be at home at 9pm. John knows this.

d. John knows that he could/may have been at home at 9pm.
Reading1: ?? *Deontic*: It is compatible with the satisfaction of the goals/with the laws for John to have been at home at 9pm. John knows this.
Reading 2: *Epistemic*: John is an amnesiac. It is compatible with the evidence that he was at home at 9pm. John knows this.

(213) finite interrogative complement

a. John knows when he must be at home.
Reading 1: *Deontic*: It is necessary for the satisfaction of some goal/by some law that John be at at home at time x. John knows

this.
Reading 2: ? *Epistemic*: John is an amnesiac. He is trying to put together his past life by looking at various pieces of evidence. From the evidence it follows that he is at home at times t_1 etc. John knows this.

b. John knows when he must have been at home.
Reading 1: ? *Deontic*: It is necessary for the satisfaction of some goal/by some law for John to have been at home at time *x*. John knows this.
Reading 2: *Epistemic*: John is an amnesiac. It follows from the evidence that he was at home at time *x*. John knows this.

c. John knows when he can/could/may be at home.
Reading 1: *Deontic*: It is compatible with the satisfaction of the goals/the law that John be at at home at time *x*. John knows this.
Reading 2: ? *Epistemic*: John is an amnesiac. He is trying to put together his past life by looking at various pieces of evidence. It is compatible with the evidence that he is at home at times t_1 etc. John knows this.

d. John knows when he can/could/may have been at home.
Reading 1: ?? *Deontic*: It is compatible with the satisfaction of the goals/the law for John to have been at home at time *x*. John knows this.
Reading 2: *Epistemic*: John is an amnesiac. It is compatible with the evidence that he was at home at time *x*. John knows this.

We see that deontic readings are easily available with *be at home* but not so easily available with *have been at home*. In order to get a deontic reading with *have been at home*, we need to set up a special context e.g. A murder has taken place. There are certain times s.t. for John to not be a suspect, he must have been at home.

Epistemic readings, on the other hand, are easily available with *have been at home* but not so easily available with *be at home*. For there to be an epistemic reading with *be at home*, *be at home* has to be interpreted generically (cf. 212a, 213a).

With infinitival questions, we find that epistemic readings are systematically absent. Even predicates like *have been at home* which favor epistemic readings only permit the otherwise less-favored deontic reading.

(214) a. John knows when to be at home.
√ Deontic: $\exists t$ John knows that being at home at time t satisfies the Goals/the Law.
* Epistemic

b. John knows when to have been at home.
√ Deontic: $\exists t$ John knows to have been at home at time t satisfies the Goals/the Law.
*Epistemic, *Circumstantial

From the absence of the epistemic readings I conclude that the modality in the infinitival question in (214) can only be deontic and not epistemic or circumstantial.

The examples we have been considering so far (212, 213, 214) were designed to bring out the epistemic reading. We will now consider examples which bring out the circumstantial reading of the modality in order to convince ourself that the circumstantial reading is indeed missing in infinitival questions.

(215) a. Stuart knows that he can solve the problem in an hour.
√ Deontic, √ Circumstantial/Ability.

b. Stuart knows how quickly he can solve the problem.
√ Deontic, √ Circumstantial/Ability.

c. Stuart knows how quickly to solve the problem.
√ Deontic, * Circumstantial/Ability.

(215) shows that while circumstantial/ability readings are available with indicative complements and finite interrogative complements, they disappear when we shift to an infinitival question (cf. 215c). Thus we reinforce our point that only deontic readings are available with infinitival questions.

We will now consider the readings available with infinitival question complements of verbs of decision like *decide*. With finite clause complements, *decide* can receive either a decision reading and a cognitive judgement reading. The decision reading corresponds to a bouletic interpretation of the embedded clause while the cognitive judgement corresponds to an epistemic interpretation of the embedded clause. Bouletic modality is a special case of Deontic Modality. As we have seen before with complements of *know*, which reading (bouletic vs. epistemic) is favored depends upon the nature of the embedded predicate. Bouletic readings are favored with *be at home* and epistemic readings with *have been at home*.

(216) a. Alfred decided yesterday that he must/may be at home at 9pm.
(√Bouletic, ???Epistemic)

b. Alfred decided yesterday that he must/may have been at home at 9pm.
(√Epistemic, ???Bouletic)

c. Alfred decided yesterday when he must/may be at home.
(√Bouletic, ???Epistemic)

d. Alfred decided yesterday when he must/may have been at home.
(√Epistemic, ???Bouletic)

e. Alfred decided yesterday to be at home at 9pm.
(√Bouletic, *Epistemic)

f. ? Alfred decided yesterday to have been at home at 9pm.
(?Bouletic, *Epistemic)

g. Alfred decided yesterday when to be at home.
(√Bouletic, *Epistemic)

h. ? Alfred decided yesterday when to have been at home.
(?Bouletic, *Epistemic)

The only readings available with infinitival complements of *decide* (whether interrogative or non-interrogative) is the decision/deontic reading. The cases with *have been at home* which otherwise favor epistemic readings are degraded and acceptable only to the extent that the deontic reading is available.

We see with *decide* that the readings that are absent with the infinitival question complement are also absent with the infinitival non-question complement. This point could not be seen with *know* because *know* did not take infinitival non-question complements.

In some cases, it seems that the infinitival modality can also be circumstantial.

(217) Magnus knows where to get gas.

It seems sufficient for Magnus to know that it is possible to get gas at place *x*. However, the option of circumstantial modality seems to disappear when we switch to a predicate like *smile, walk, talk* etc. Assume that Hafdis lives in a world where there is interplanetary travel and where it is possible for humans to fly on some planets but not all. In addition, on some planets even though it is physically possible for humans to fly, it is not permitted by the planetary administration.

(218) a. Hafdis knows where she can/may fly.
Deontic (where she is permitted), Circumstantial (where it is possible)

b. Hafdis knows where to fly.
Deontic/Bouletic (where it is appropriate given laws/desires), *Circumstantial (where it is possible)

Since the existence of the circumstantial modality reading in (217) is somewhat restricted, we should consider the environments which make such a reading available more carefully and not assume it to be a generally available option. In Sect. 3.1, we discuss why the circumstantial modality reading is available in (217) and why it is not available more generally.

2.2. The force of infinitival modality

If we try to paraphrase an infinitival question with a finite question, we find that the paraphrase involves a modal word. The quantificational force of the modal word that we choose seems to vary depending upon a variety of factors such as the embedding predicate, the extracted element, the predicate in the infinitival question, and the context. The modal that we choose can be *could*, *should*, or *would*. In this section, a description is provided of some of the factors which favor one reading or the other. A discussion of why some environments only permit a certain modal force or tend to favor a certain modal force can be found in Sect. 3.

2.2.1. *The effect of the* wh-*word*

A paraphrase with *can/could* seems appropriate when the extracted element is *where* and *how*.

(219) a. Magnus knows where to get gas.
≈ Magnus knows where he/one can get gas.

b. Magnus knows how to solve this problem.
≈ Magnus knows how one can solve this problem.

With other *wh*-words like *who*, *whether*, *when*, the temporal use of *where*, *how many* N, *how*-Adj/Adv, and selected *how*, paraphrases with *should* seem more appropriate than paraphrases with *could*.

(220) a. Magnus knows who to invite to the party.
b. Magnus knows whether to invite Siggi to the party.
c. Magnus knows when to call me up.
d. Magnus knows where to interrupt.
e. Magnus knows how many people to invite to the party.
f. Magnus knows how quickly to solve the problem.
g. Magnus knows how to word the letter.

Should paraphrases are also more appropriate than paraphrases with *could* with *which* phrases, singular or plural.

(221) a. Magnus knows which book to read for tomorrow's class.
b. Magnus knows which gas stations to get gas from. (compare with 219a)
c. ?Magnus knows in which ways to solve the problem. (compare with 219b)

(221c) is a bit awkward due to the pied-piping, but the contrast with (219b) is quite sharp.

2.2.2. *The effect of the embedding predicate*

The choice of the embedding predicate also plays a role in determining what we feel the force/flavor of the infinitival modality is.

(222) a. Hafdis and Ásta know where to get gas.
b. Hafdis and Ásta agreed on where to get gas.

The intuition is that for (222b), in contrast to (222a), *should* or *would* is a more appropriate paraphrase than *could*.

If we change the matrix predicate from *know* to *decide*, a *would* paraphrase seems to become the favored one.

(223) a. Ásta knows where to get gas.
b. Ásta has decided where to get gas.

In (224), we go through the list of predicates that take infinitival question complements (organized according to Karttunen's classification) to see the influence of the embedding predicate in determining what we feel the force/flavor of the infinitival modality is. The modality that seems most suitable in a paraphrase involving an overt modal is indicated in each case below the relevant example.

(224) a. Ásta knows where to get gas.
(verb of retaining knowledge, *could*)

b. Ásta discovered/figured out where to get gas.
(verb of acquiring knowledge, *could*)

c. Ásta decided where to get gas.
(decision verb, *would*)

d. Ásta is unsure about where to get gas.
(opinion verb, *should*)

e. Ásta told Hafdis where to get gas.
(verb of one-way communication, *should*)

f. Ásta and Hafdis argued over/talked about where to get gas.
(verb of two-way communication, *should*)

g. Ásta studied/is considering where to get gas.
(verb of cogitation, *should*)

h. Ásta asked Hafdis/wondered where to get gas.
(inquisitive verb, *should*)

The discussion in this chapter will be limited to infinitival question complements of *know*. A study of the interaction between the semantics of embedding verbs other than *know* and the semantic of infinitival questions is left for future research.

2.2.3. *The effect of the infinitival question predicate*

If the extracted element is *where*, we find that *could* paraphrases are appropriate only if the embedded predicate is something like *find, get, obtain* etc.

(225) a. Ásta knows where to get gas.

b. Ásta knows where to eat lunch.

In (225b), a *should* paraphrase seems more appropriate than a *could* one. However, consider (226), which disfavors a *should* paraphrase.

(226) Ásta knows where to eat well. (where *to eat well* means to eat good food, not to eat heartily)

More generally, it seems that *could* paraphrases are appropriate for predicates which imply attainment of a goal. In (225a), the goal seems to be to get gas and in (226), to eat well. Satisfaction of the infinitival question predicate entails satisfaction of the goals. In contrast, in (225b) where a *could* paraphrase seems inappropriate, the goal is not to merely eat lunch.

2.2.4. *The effect of the context*

We find that in most cases where the most appropriate paraphrase seems to be a *could* paraphrase, if we modify the context a non-*could* paraphrase becomes available.

(227) a. Magnus knows where to get gas.
(context: we want gas but we are boycotting gas-stations owned by the Radical Right)

b. Magnus knows how to solve this problem.
(context: The goal is to solve the problem but without violating any social norms)

In general the intuition seems to be that if just any alternative will do, a *could* paraphrase seems appropriate. Otherwise a *should* paraphrase seems more natural.

3. The modality in infinitival questions

My basic proposal is that despite the seeming variability in the modality of infinitival questions, all infinitival questions involve the same modality. The apparent variation between the *could*, *should*, and *would* readings comes from properties of the context and the semantics of the embedding verb among other things. It will be shown that it is not necessary to postulate the existence of different underlying modals to explain the apparent variability in the interpretation of infinitival questions.

We have noted before that the modality in infinitival questions can only be deontic/bouletic and not epistemic. Modality has been analyzed along the following three dimensions: the force of the modality, the modal base, and an ordering source. (cf. Kratzer (1981, 1991)). In the case of deontic modality, the modal base is a circumstantial one. A circumstantial modal base assigns to every possible world a set of propositions that are true in that world. Further, for a deontic modality, the ordering source is one which orders worlds according to how closely they approach the ideal established by the law/the goals/what is moral/what we want (bouletic) etc.

Let us begin with a consideration of the definitions of the deontic modalities $\Box_D$ and $\Diamond_D$. The circumstantial modal base provides a restriction on the worlds that we consider and out of these worlds the deontic ordering source tells us in which worlds the goals are met/ the laws are upheld etc.[55]

(228) *CMB* is the circumstantial modal base, *DOS* is the deontic ordering source. *CMB* is a function from worlds to sets of propositions; *DOS* is a function from sets of worlds to sets of worlds - when applied to a set of worlds *A*, *DOS* returns a subset of *A*, which most closely approximates the ideal established by the ordering source.

a. $[\![\Box_D(p)]\!]^w = 1 \iff \forall w'[w' \in DOS(\bigcap CMB(w)) \rightarrow p(w')]$

b. $[\![\Diamond_D(p)]\!]^w = 1 \iff \exists w'[w' \in DOS(\bigcap CMB(w)) \wedge p(w')]$

We can think of the deontic ordering source as specifying the goals that have to be met. The above definitions state that the proposition p holds in every/some world which is picked out by the circumstantial modal base and the deontic ordering base. If we think of p as an action that we have to execute, if $\Box_D(p)$ is true, we know that the action p is executed in all worlds picked out by the circumstantial modal base where the goals are met. However, it does not tell us that executing the action p will lead to the satisfaction of the goals. This point can be seen in the following example.

(229) a. Situation 1: My goal is to become popular. One way to become popular is to invite Daniel, Hafdis, and Biggi.

b. Situation 2: My goal is to become popular. The way to become popular is to invite Daniel, Hafdis, and Biggi.

In situation 1, the proposition $\Diamond_D(invite(d))$ and in situation 2, the proposition $\Box_D(invite(d))$ is true. Yet execution of the action *invite*(d) alone is not sufficient for the satisfaction of the goals. In situation 1, inviting Daniel, Hafdis and Biggi will lead to satisfaction of the goals though this may not be the only way of satisfying the goals. In situation 2, inviting Daniel, Hafdis and Biggi will also lead to satisfaction of the goals. It is also the only way to satisfy the goals. I will define versions of deontic necessity and possibility which will only pick out actions execution of which satisfies the goals. First an informal definition:

(230) a. $[\![\Box_{D,\rightarrow}(p)]\!]^w = 1 \iff$
(i) Action p is necessary. (ii) Execution of p satisfies Goals. (Action p is the unique way to satisfy the goals)

b. $[\![\Diamond_{D,\rightarrow}(p)]\!]^w = 1 \iff$
(i) Action p is possible. (ii) Execution of p satisfies Goals. (Action p is one way to satisfy the goals)

We will attempt to formalize the following intuitions.

(231) Hafdis knows [who to talk to at the party]. (Goals: popularity) Talking to Magnus, Herb, and Penna can deliver popularity. Talking to Daniel, Stefan-Árni, and Baldur can also deliver popularity.

Firstly it is insufficient for the truth of (231) for Hafdis to merely know one of the following propositions: (talking to Magnus can satisfy goals; talking to Herb can satisfy goals; talking to Penna can satisfy goals; etc.). Secondly it is sufficient for the truth of (231) that Hafdis know that one of the following actions (talking to Magnus, Herb, and Penna; talking to Daniel, Stefan-Árni, and Baldur) will satisfy the goals. So while it is not necessary to know all the answers, it seems necessary to know *enough*. Intuitively, one knows *enough* if what one knows will help one to achieve one's goals. This notion of *enough* will be built into the definition of the infinitival modality.

We can now spell out the denotations of the enriched modalities in (230) (cf. 232).

(232) From this point onwards, we will use *Goal* to refer to the function $\lambda w.DOS(\bigcap CMB(w))$. *DOS*, the deontic ordering source, is a function from sets of worlds to sets of worlds, and *Goal* from worlds to propositions/sets of worlds.
We will also use *Rel* to refer to the function $\lambda w.\bigcap CMB(w)$. *CMB*, the circumstantial modal base, is a function from worlds to sets of propositions, and *Rel* (=Relevant) from worlds to propositions/sets of worlds.
The set of worlds picked out by *Goal* is always a subset of the set of worlds picked out by *Rel* i.e. $\forall w[Goal(w) \rightarrow Rel(w)]$.

$$\text{a. } [\![\Box_{D,\rightarrow}(p)]\!]^w = 1 \iff \forall w'[w' \in Goal(w) \rightarrow p(w')] \wedge \forall w'[[w' \in Rel(w) \wedge p(w')] \rightarrow w' \in Goal(w)] \iff \forall w'[w' \in Rel(w) \rightarrow [w' \in Goal(w) \iff p(w')]]$$

$$\text{b. } [\![\Diamond_{D,\rightarrow}(p)]\!]^w = 1 \iff \exists w'[w' \in Goal(w) \wedge p(w')] \wedge \forall w'[[w' \in Rel(w) \wedge p(w')] \rightarrow w' \in Goal(w)]$$

The definition in (232a) says that if we restrict ourself to the worlds in the modal base, *p* is equivalent to the propositional content of the goals in the current world. This modal operator is basically one that checks for identity between propositions. Since it is not clear where such an operator would be useful, we will not use this strengthened modality for infinitival questions.

Let us now look at $\Diamond_{D,\rightarrow}$ in the context of (231). According to our proposal, it is this modality that appears in infinitival questions. With this modality in hand we can compute the denotation of the infinitival question in (231) (cf. 233). I assume a Hamblin-Karttunen semantics for questions , where a question denotes a set of propositions (cf. Hamblin 1973, Karttunen 1977).

(233) $[\![[who_i\ [\Diamond_{D,\rightarrow}[PRO\ to\ talk\ to\ t_i]]]]\!] =$
$\lambda p \exists x[Person(x) \wedge p = \Diamond_{D,\rightarrow}(talk(hafdis,x))]$
Since *know* is factive, we will only consider the true answers
$= \{\Diamond_{D,\rightarrow}(talk(hafdis,M+H+P)), \Diamond_{D,\rightarrow}(talk(hafdis,D+S+B))\}$

Note that in order to generate propositions of the form $\Diamond_{D,\rightarrow}(talk(hafdis,M+H+P))$, we will need to allow the variable bound by *who* to range over plural objects. This is unproblematic since *who* is unmarked for number. If the *wh*-phrase is marked for number e.g. *which person*, the question denotation permits only propositions involving singular objects in the trace position as expected. Note also that since we are using pluralities, we have to have a way to pick out the *most informative* answers. In addition to the elements shown in (233), the question denotation in (233) will contain many more elements (cf. 234).

(234) $\{\Diamond_{D,\rightarrow}(talk(hafdis,M+H+P)), \Diamond_{D,\rightarrow}(talk(hafdis,D+S+B)),$
$\Diamond_{D,\rightarrow}(talk(hafdis,M+H+P+A_1)), \Diamond_{D,\rightarrow}(talk(hafdis,M+H+P+A_2)), \Diamond_{D,\rightarrow}(talk(hafdis,M+H+P+A_1+A_2)),$
$\Diamond_{D,\rightarrow}(talk(hafdis,D+S+B+A_1)), \Diamond_{D,\rightarrow}(talk(hafdis,D+S+B+A_2)), \Diamond_{D,\rightarrow}(talk(hafdis,D+S+B+A_1+A_2))\}$
where A_1 and A_2 are two arbitrary individuals.

One way to pick out the most informative answers is to pick up the smallest subset of the question denotation that entails everything else in the question denotation.[56]

Because of the way the modality is defined, the set of true propositions in the question denotation does not include propositions like $\Diamond_{D,\rightarrow}(talk(hafdis,M)), \Diamond_{D,\rightarrow}(talk(hafdis,H)), \Diamond_{D,\rightarrow}(talk(hafdis,M+H))$ etc. Talking to M, or to H, or to both does not lead to satisfaction of the goals. This captures Intuition 1.[57]

The second component of the proposal is that there is no exhaustivity in the interpretation of infinitival questions, not even weak exhaustivity. So in order to know an infinitival question, it is sufficient to know any one of the propositions in its denotation. Hence it is sufficient for Hafdis to know either $\Diamond_{D,\rightarrow}(talk(hafdis,M+H+P))$ or $\Diamond_{D,\rightarrow}(talk(hafdis,D+S+B))$. It is not necessary to know both.[58] The absence of weak exhaustivity in infinitival questions is in contrast with finite questions where there is weak exhaustivity and arguably strong exhaustivity (cf. 235).

(235) Snorri knows who Ásta likes.
True answers to *who Ásta likes* = { Ásta likes Anna, Ásta likes Sjón, Ásta likes Hafdis }

Suppose that Snorri only knows that Ásta likes Sjón but does not know that Ásta likes Anna and Hafdis too. Then it is not true that (235) is true. The requirement that in order to know the answer to a finite question one has to know all the true propositions in the question denotation is called the requirement of *weak exhaustivity.*[59]

3.1. *Could* readings

We noted before that there are certain cases where it seems that the modality in an infinitival question can be paraphrased by *could* while in most other cases a *could* paraphrase is not appropriate. These examples are repeated in (236).

(236) a. Biggi knows where to get gas.
($\approx$ Biggi knows where he/one can get gas)

b. Siggi knows how to prove this theorem.
($\approx$ Biggi knows how he/one can prove this theorem)

If the modality indeed stays uniform between the cases discussed in the last section, we need to explain why it is only in cases like (236) that a *could* paraphrase is acceptable and not in cases like (231).

The difference between (236), where a *could* paraphrase is available, and (231), where a *could* paraphrase is not follows from the context. In particular, this difference follows from the difference in the goals typically associated with (236) and (231).

We saw that the goal associated with the infinitival question in (231) was not achieved by the execution of just any arbitrary deontically possible action. In (231), the goal was to become popular and this goal was not met by the execution of just any deontically possible action. For example, while talking to Magnus was a deontically possible action, its execution would not lead to satisfaction of goals. In other words, the second conjunct in the definition of the modality did some work for us.

Now let us consider the cases in (236). In (236a) the goal is to have gas and in (236b) the goal is to have a proof of the theorem/have the theorem proved. The definition of $\Diamond_{D,\rightarrow}$ is repeated in (237).

(237) $$[\![\Diamond_{D,\rightarrow}(p)]\!]^w = 1 \iff \exists w'[w' \in Goal(w) \wedge p(w')] \\ \wedge \forall w'[[w' \in Rel(w) \wedge p(w')] \rightarrow w' \in Goal(w)]$$

Suppose that the locations where it is possible to get gas are $L_1,\ldots,L_n$. Getting gas at any of these locations would lead to one having gas. Therefore the second conjunct in (237) is trivially satisfied since left hand side of the material implication will always entail the right hand side.

This is also the case with (237b). Suppose that the methods by which it is possible to prove this theorem are $M_1,\ldots,M_n$. Proving the theorem by any of these ways would lead to one having a proof of the theorem/having the theorem proved. Again the second conjunct in (237) is trivially satisfied since left hand side of the material implication will always entail the right hand side.

As a result of the trivial satisfaction of the second conjunct, in the special cases of *how to* and *where to*, $\Diamond_{D,\rightarrow}$ reduces to plain deontic possibility.

(238) Special case of $\Diamond_{D,\rightarrow}$
If $\forall w'[[w' \in Rel(w) \wedge p(w')] \rightarrow w' \in Goal(w)]$ is trivially true, $\Diamond_{D,\rightarrow}$ becomes equivalent to ordinary deontic possibility:
$$[\![\Diamond_{D,\rightarrow}(p)]\!]^w = 1 \iff \exists w'[w' \in Goal(w) \wedge p(w')] \iff \Diamond_D(p)$$

Since the satisfaction of the goals is entailed by p, the deontic component can be eliminated. We are now left with a pure circumstantial possibility.

(239) Special case of $\Diamond_{D,\rightarrow}$
$[\![\Diamond_{D,\rightarrow}(p)]\!]^w = 1 \iff \exists w'[w' \in Goal(w) \wedge p(w')] \iff \Diamond_D(p)$
However, $\forall w'[[w' \in Rel(w) \wedge p(w')] \Longrightarrow w' \in Goal(w)]$. Therefore
$\exists w'[w' \in Goal(w) \wedge p(w')] \iff \exists w'[w' \in Rel(w) \wedge p(w')]$
$\Diamond_D(p) \iff \Diamond_C(p)$

This seems to be the right result. We had noted in Sect. 2.1 that while in general the infinitival modality could only be interpreted as a deontic/bouletic modality, there were certain environments where a circumstantial reading seemed appropriate. It was also noted that an epistemic possibility reading was never available. We have now provided an explanation for why a circumstantial reading emerges in the environments under discussion. Further, we also have an explanation for why we get a circumstantial possibility reading and not an epistemic possibility reading. The explanation has two parts: firstly the pure possibility reading emerges out of a special case of $\Diamond_{D,\rightarrow}$, which is a deontic modality, and secondly deontic ordering sources and epistemic modal bases do not combine. (cf. Kratzer 1991, pg. 649). Consequently, the modal base cannot be an epistemic one and an epistemic possibility reading is not possible.

We can force the *could* reading by incorporating the goals into the body of the infinitival question. When by doing so, we make the second conjunct trivially true, we get a *could* reading.

(240) Siggi knows where to eat well.
$\approx$ Siggi knows where he/one can eat well.
$\neq$ Siggi knows where he/one should eat well.

The goal in (240) is to to have good food.[60] Suppose the establishments where it is possible to eat well are $E_1, \ldots, E_n$. Eating at any one of them is enough to satisfy the goal of having good food. As a result, the $\Diamond_{D,\rightarrow}$ reduces to plain possibility $\Diamond_D$.

It should be noted that *should* readings are always available: in general wherever *could* paraphrases are available, a *should* paraphrase can also be had by manipulation of the context. If we put more content into the goals associated with a particular infinitival question where a *could* paraphrase is available, we get a *should* reading.

(241) Baldur knows where to get gas.

Context: It is socially inappropriate to get gas from gas-stations owned by the Morally Repugnant Elite.
Goal: to have gas, without violating social norms.

In (241), the intuition seems to be that a *could* paraphrase is not appropriate.

3.2. The role played by the context

The semantics presented for infinitival questions in the previous section is somewhat simplistic. Let us examine the meaning of $\Diamond_{D,\rightarrow}$ more closely.

(242) $[\![\Diamond_{D,\rightarrow}(p)]\!]^{w} = 1 \iff$
$$\exists w'[w' \in Goal(w) \wedge p(w')] \wedge$$
$$\forall w'[[w' \in Rel(w) \wedge p(w')] \rightarrow w' \in Goal(w)]$$

According to (242), for $\Diamond_{D,\rightarrow}(p)$ to be true in a world w, in every relevant world where the action corresponding to p is executed the goals must be met. Consider this definition in the context of (231), repeated here as (243).

(243) Hafdis knows [who to talk to at the party]. (Goals: popularity)
Talking to Magnus, Herb, and Penna can deliver popularity. Talking to Daniel, Stefan-Árni, and Baldur can also deliver popularity.
$[\![[who_i\ [\Diamond_{D,\rightarrow}[PRO\ to\ talk\ to\ t_i]]]]\!] =$
$\lambda p \exists x[Person(x) \wedge p = \Diamond_{D,\rightarrow}(talk(hafdis,x))]$
Since *know* is factive, we will only consider the true answers
$= \{\Diamond_{D,\rightarrow}(talk(hafdis,M+H+P)), \Diamond_{D,\rightarrow}(talk(hafdis,D+S+B))\}$

For $\Diamond_{D,\rightarrow}(talk(hafdis,M+H+P))$ to be true, it is necessary that in all worlds where Hafdis talks to Magnus, Herb, and Penna, she becomes popular. This is too strong. Suppose Hafdis does talk to Magnus, Herb, and Penna but also makes retching noises while talking to them, it is unlikely that she will become popular. Clearly there are many worlds where Hafdis does talk to Magnus, Herb, and Penna but for some reason (cf. getting violently drunk, breaking furniture, telling inappropriate jokes, assaulting the guests etc.) does not become popular. Still the intuition about (243) is that it is enough to know that talking to Magnus, Herb, and Penna will lead to popularity. The fact that making retching noises will stop Hafdis from achieving her goals does not seem to be relevant for determining the truth of (243).

In other worlds, it seems that we do not require all relevant worlds where the action p (= talking to Magnus, Herb, and Penna) is executed to be worlds where the goals are satisfied. Very informally, we first remove the offensive worlds where Hafdis breaks the furniture, she makes retching noises etc. Once these offending worlds are removed from the domain of quantification, in all the other relevant worlds where p is satisfied the goals must be satisfied. This can be expressed as follows using a contextual restriction C on the domain of quantification of the universal quantifier in (242) (cf. 244).

(244) $[\![\Diamond_{D,\rightarrow}(p)]\!]^{w} = 1 \iff$
$$\exists w'[w' \in Goal(w) \wedge p(w')] \wedge$$
$$\forall w'[[w' \in C \wedge p(w')] \rightarrow w' \in Goal(w)]$$
Constraint on C: $C \subseteq Rel(w)$

We need to make precise how the contextual variable C is set. We have the intuition that the existence of worlds where Hafdis breaks furniture and makes retching noises is somehow irrelevant for the purposes of the infinitival question and so we assumed that they could excluded from the domain of quantification. What makes a world relevant/irrelevant needs to be made precise.

Now recall that one intuition regarding (243) that we wanted to explain was why $\Diamond_{D,\rightarrow}(talk(hafdis,M))$ did not count as a sufficiently complete answer. Our explanation was that talking to Magnus did not lead to satisfaction of goals in all the relevant worlds (e.g. the worlds where Hafdis talks to Magnus but not Herb or Penna). Therefore $\Diamond_{D,\rightarrow}(talk(hafdis,M))$ was false. But now that we are not considering *all* worlds where the action p is executed but only a subset, what stops us from considering out of the worlds where Hafdis talks to Magnus, just those worlds where Hafdis also talks to Herb and Penna. This work should be done by the mechanism which sets the value of the contextual restriction C. The mechanism should be such that worlds where Hafdis breaks furniture or makes retching noises are somehow excluded from C but not worlds where Hafdis does not talk to Penna and world where Hafdis does not talk to Herb.

How can we distinguish between worlds where Hafdis breaks furniture and worlds where Hafdis does not talk to Penna in the context of (243)? The intuition that will be pursued here is that the worlds where Hafdis breaks furniture are irrelevant in the context of (243) because they are not part of the alternatives evoked by the question meaning. The worlds where Hafdis does not talk to Penna are relevant because they are part of the alternatives evoked by the question meaning.

To make the notion of alternatives evoked by a question precise, I will use the notion of question denotation developed by Groenendijk & Stokhof (1984) and Higginbotham (1993). The question denotation is taken to denote a partition of the logical space. Each cell of the partition represents one way the world can be with respect to the meaning of the question. For example, consider the question in (245).

(245) Who did Hafdis talk to?
assume domain of individuals consists of Magnus, Herb, and Penna

The partition corresponding to (245) is in (246).

(246) a. talked(hafdis,m) $\wedge$ talked(hafdis,h) $\wedge$ talked(hafdis,p)
b. $\neg$ talked(hafdis,m) $\wedge$ talked(hafdis,h) $\wedge$ talked(hafdis,p)
c. talked(hafdis,m) $\wedge\neg$ talked(hafdis,h) $\wedge$ talked(hafdis,p)
d. talked(hafdis,m) $\wedge$ talked(hafdis,h) $\wedge\neg$ talked(hafdis,p)
e. $\neg$ talked(hafdis,m) $\wedge\neg$ talked(hafdis,h) $\wedge$ talked(hafdis,p)
f. $\neg$ talked(hafdis,m) $\wedge$ talked(hafdis,h) $\wedge\neg$ talked(hafdis,p)
g. talked(hafdis,m) $\wedge\neg$ talked(hafdis,h) $\wedge\neg$ talked(hafdis,p)
h. $\neg$ talked(hafdis,m) $\wedge\neg$ talked(hafdis,h) $\wedge\neg$ talked(hafdis,p)

Whether Hafdis breaks the furniture or makes objectionable noises in a particular world w does not influence which cell of the partition that world w goes in to. On the other hand whether Hafdis talks to a certain person, say Penna, or not in a particular world w does influence which cell of the partition the world w goes into.

I propose that the question meaning imposes the following constraint on C.

(247) Let $\mathscr{P}$ be the partition associated with the question Q. $\mathscr{P}$ is a set of propositions. Assume $\mathscr{P} = \{qa_1, \ldots, qa_n\}$.
When evaluating $\Diamond_{D,\rightarrow}$, C must satisfy the following conditions:
• $\forall qa[[qa \in \mathscr{P}] \Longrightarrow \exists w[w \in qa \wedge w \in C]]$
This constraint says that C must contain one representative world from each cell of the partition. (Positive Constraint on C)
• As long as the positive constraint is satisfied, C is chosen in such a way that only those worlds are included where everything that is conducive to the attainment of the goals is done. (Constraint of Charity on C)

With the modified semantics of $\Diamond_{D,\rightarrow}(p)$, we have a semantics that is neither too weak (it does not ignore worlds where Hafdis doesn't talk to Penna) nor too strong (it ignores worlds where Hafdis makes retching noises). For concreteness, let us see how in the context of (243), (248a) comes out false and how (248b) comes out true.

(248) a. $\Diamond_{D,\rightarrow}(talk(hafdis,m)) \iff$
$\exists w'[w' \in Goal(w) \wedge talk(hafdis,m,w')] \wedge$
$\forall w'[[w' \in C \wedge talk(hafdis,m,w')] \rightarrow w' \in Goal(w)]$

By the positive constraint on C, C contains a world where Hafdis does not talk to Herb. In this world, she does not become popular i.e. the goals are not met and therefore $\Diamond_{D,\rightarrow}(talk(hafdis,m))$ is false in the world of evaluation.

b. $\Diamond_{D,\rightarrow}(talk(hafdis,m+h+p)) \iff$
$\exists w'[w' \in Goal(w) \wedge talk(hafdis,m+h+p,w')] \wedge$
$\forall w'[[w' \in C \wedge talk(hafdis,m+h+p,w')] \rightarrow w' \in Goal(w)]$

There may be worlds where Hafdis makes retching noises in addition to talking to Magnus, Herb, and Penna and therefore does not become popular. However, due to the charity constraint on C, this world is not included in C. There is a world in the cell of the partition that corresponds to Hafdis talking to Magnus, Herb, and Penna where Hafdis does become popular. This is enough to make $\Diamond_{D,\rightarrow}(talk(hafdis,m+h+p))$ true in the world of evaluation.

3.2.1. Whether *questions*

The introduction of the contextual restriction C on the domain of the universal quantifier in the semantics of $\Diamond_{D,\rightarrow}$ has the welcome side-effect of enabling us to handle infinitival *whether* questions. Consider (249) in the following context: the goal is to become popular. Hafdis can become popular by talking to Magnus, Herb, and Penna.

(249) Hafdis knows [whether to talk to Magnus].
$[\![whether\ to\ talk\ to\ Magnus]\!] =$
$\{\Diamond_{D,\rightarrow}(talk(hafdis,m)), \neg\Diamond_{D,\rightarrow}(talk(hafdis,m))\}$

Since *know* is factive[61], Hafdis knows the true element in the denotation of the infinitival question. Our intuition about (249) is that in this context, the positive element of the question denotation, $\Diamond_{D,\rightarrow}(talk(hafdis,m))$, is true. However, given the unrestricted semantics for $\Diamond_{D,\rightarrow}$, $\Diamond_{D,\rightarrow}(talk(hafdis,m))$ is false. It is made false because of the existence of relevant worlds where Hafdis talks to Magnus but not to Herb or Penna and does not become popular making the second conjunct false.

This problem does not arise with the contextual restriction augmented semantics for $\Diamond_{D,\rightarrow}$. The domain of quantification is sensitive to the form of the question indirectly by being constrained by the partition created by the question. The partition created by a *whether* question only has two elements: a positive element and a negative element. In the case at hand, at least one world from the negative element in the partition does enter C because of the positive constraint on C. In such a world, $talk(hafdis,m)$ is false.

(250) $\Diamond_{D,\rightarrow}(talk(hafdis,m)) \iff$
$$\exists w'[w' \in Goal(w) \wedge talk(hafdis,m,w')] \wedge$$
$$\forall w'[[w' \in C \wedge talk(hafdis,m,w')] \rightarrow w' \in Goal(w)]$$

The worlds in C where $talk(hafdis,m)$ is false do not matter because the falsehood of $talk(hafdis,m)$ makes the material implication true for such worlds. C also contains worlds where $talk(hafdis,m)$ is true. By the charity constraint on C, we only include worlds where everything conducive to the satisfaction of the goals takes place. So we include the worlds where Hafdis talks to Magnus, Herb and Penna, does not do anything inappropriate, and becomes popular. However, we do not include the worlds where Hafdis only talks to Magnus, or the worlds where Hafdis talks to Magnus, Herb, and Penna but also behave inappropriately and does not become popular. Hence the second conjunct in the definition of $\Diamond_{D,\rightarrow}(talk(hafdis,m))$ is true. The first conjunct in the definition of $\Diamond_{D,\rightarrow}(talk(hafdis,m))$ is true because there is a relevant world where Hafdis talks to Magnus and becomes popular. This makes $\Diamond_{D,\rightarrow}(talk(hafdis,m))$ a true answer to the infinitival question *whether to talk to Magnus* in (249).

3.2.2. *A prediction: the behavior of* what to do *questions*

In the context of (243), we permitted exclusion of 'irrelevant' activities like breaking the furniture and making retching noises by proposing a constraint on what the contextual restriction had to contain. This constraint allowed us to ignore 'irrelevant' activities when constructing the contextual restriction because such activities were not an essential part of the question partition. The prediction that we now make is that if we have a question where such 'irrelevant' activities are part of the question partition, then they will no longer be 'irrelevant'. We will not be able to ignore them. This prediction is met (cf. 251).

(251) Hafdis knows [what to do at the party].

The partition evoked by the question *what to do at the party* is shown in (252).

(252) Assume for simplicity that the activities in the domain are: talking to Magnus, talking to Herb, and breaking furniture. Assume also that the goals are to become popular and that these goals are met by talking to Magnus and Herb and not breaking furniture.

a. $talk(hafdis,m) \wedge talk(hafdis,h) \wedge break(hafdis,f)$ (-Goal)

b. $\neg talk(hafdis,m) \wedge talk(hafdis,h) \wedge break(hafdis,f)$ (-Goal)

c. $talk(hafdis,m) \wedge \neg talk(hafdis,h) \wedge break(hafdis,f)$ (-Goal)

d. $talk(hafdis,m) \wedge talk(hafdis,h) \wedge \neg break(hafdis,f)$ (+Goal)

e. $\neg talk(hafdis,m) \wedge \neg talk(hafdis,h) \wedge break(hafdis,f)$ (-Goal)

f. $\neg talk(hafdis,m) \wedge talk(hafdis,h) \wedge \neg break(hafdis,f)$ (-Goal)

g. $talk(hafdis,m) \wedge \neg talk(hafdis,h) \wedge \neg break(hafdis,f)$ (-Goal)

h. $\neg talk(hafdis,m) \wedge \neg talk(hafdis,h) \wedge \neg break(hafdis,f)$ (-Goal)

The intuition about (251) is that it is not enough that Hafdis know that talking to Magnus and Herb will satisfy her goals. She has to know that talking to Magnus and Herb and not breaking furniture will satisfy her goals. This intuition is captured.

$\Diamond_{D,\rightarrow}(talk(hafdis,m+h))$ is false - there are two cells of the partition in (252) where $talk(hafdis,m+h)$ is true. The goals are however only met in (252d) and not in (252a) making $\Diamond_{D,\rightarrow}(talk(hafdis,m+h))$ false. In contrast, $\Diamond_{D,\rightarrow}(talk(hafdis,m+h) \wedge \neg break(hafdis,f))$ is true. There is only one cell in the partition in (252) where $talk(hafdis,m+h) \wedge \neg break(hafdis,f)$ is true and in this cell (= 252d), the goals are met.

3.3. Restrictions on goals

The infinitival modality $\Diamond_{D,\rightarrow}$ imposes a stronger relationship between its propositional argument p and the Goals $\mathcal{G}$ than is imposed by $\Diamond_D$. Cf. (253a-c), which involve an overt *can*. The goal with an overt *can* does not have to be 'caused' by the propositional argument of *can*. It just has to be compatible with it. For example in (253a), the goals can be to pass the exam, something that is not caused by skipping classes, or the goals could also be to avoid getting a headache, something that can be caused by skipping certain classes.

(253) a. Siggi knows which classes he can skip. (and still pass the exam)/ (# in order to pass the exam)/ (in order to avoid getting a headache)

b. Siggi knows which questions he can skip. (and still pass the exam)/ (# in order to pass the exam)/ (in order to save time)

c. Siggi knows which questions he can screw up on. (and still pass the exam)/ (# in order to pass the exam)/ (?in order to transmit secret messages to the people grading his exam)

d. Siggi knows which classes to skip.

e. Siggi knows which questions to skip.

f. Siggi knows which questions to screw up on.

In contrast to (253a-c), (253d-f), which involve the infinitival modality $\Diamond_{D,\rightarrow}$, only allow goals that are, in some sense, caused by the propositional argument of $\Diamond_{D,\rightarrow}$. For example (253d) only permits the goals to be 'to avoid getting a headache' and not 'to pass the exam'. We will build this causal relationship into the definition of $\Diamond_{D,\rightarrow}$ by replacing the material implication in the second conjunct by a causal implication.

Examples like (254) throw further light on the relationship between the propositional argument of $\Diamond_{D,\rightarrow}$ and the goals.

(254) Biggo knows what to read.
Goal: to get gas.

It is clear that if we make the plausible assumption that what Biggo reads has no effect on getting gas, (254) cannot be used with the indicated goals. Why this is so also seems intuitively clear. We feel that this is so because there does not seem to be any non-trivial relationship between Biggo's reading something and Biggo's getting gas. In a context where one can get gas only by reading certain texts, it does seem possible to use (254) with the goals indicated. This relationship between the propositional argument of $\Diamond_{D,\rightarrow}$ and the goals is not captured by anything in the semantics of $\Diamond_{D,\rightarrow}$. To see why, consider that for (254) to be true, Biggo has to know a true member of the infinitival question denotation. So for some book k, if $\Diamond_{D,\rightarrow}(read(biggo,k))$ is true and Biggo knows that $\Diamond_{D,\rightarrow}(read(biggo,k))$, then (254) is true. In general, $\Diamond_{D,\rightarrow}(read(biggo,k))$ will be true for any book k that Biggo reads.

(255) $\Diamond_{D,\rightarrow}(read(biggo,k)) \iff$
$\exists w'[w' \in Goal(w) \wedge read(biggo,k,w')] \wedge$
(true: there is a relevant world where Biggo reads k and gets gas)
$\forall w'[[w' \in C \wedge read(biggo,k,w')] \rightarrow w' \in Goal(w)]$
(true: By our charity constraint on C, in each world in C where Biggo reads k, Biggo gets gas.)

Our semantics for $\Diamond_{D,\rightarrow}$ does not predict the oddness of (254) with the indicated goals so far.

We will build in the relationship of causation that holds between the propositional argument of $\Diamond_{D,\rightarrow}$ and the goals into the semantics of $\Diamond_{D,\rightarrow}$. We do not provide a formal analysis of causation here, simply assuming that one exists (call it $\rightarrow_c$) and use it to replace the material implication in the second conjunct of the definition of $\Diamond_{D,\rightarrow}$. Once we have this revised definition of $\Diamond_{D,\rightarrow}$, we will be able to exclude cases like (254) because of the absence of a relationship of causation between the propositional argument of $\Diamond_{D,\rightarrow}$ and the goals. The results obtained so far with the material implication $\Diamond_{D,\rightarrow}$ will also hold with the causative implication $\Diamond_{D,\rightarrow}$.

3.4. *Should* readings

We have seen why in some cases it is possible to paraphrase the infinitival modality with a possibility modal. We have also seen why in other cases it is not possible to paraphrase the infinitival modality with a possibility modal. In the previous sections, certain environments have been discussed where *should* paraphrases are appropriate. The appropriateness of *should* paraphrases in these environments will be explained in this section.

It will be shown how the semantics of $\Diamond_{D,\rightarrow}$ combined with certain other factors is equivalent to a *should* reading. These factors include explicit singular marking on the *wh*-phrase (Sect. 3.4.1), and the influence of the context (Sect. 3.4.2).

There are also other environments where the infinitival modality is not equivalent to a *could* or a *should*. These environments still seem to permit a *should* paraphrase. I argue that this is so because modal questions with *should* permit 'non-exhaustive' interpretations, and the infinitival modality is equivalent to the 'non-exhaustive' interpretation of a modal question with *should* (Sect. 3.4.3).

3.4.1. *Singular* wh-*phrases*

If the *wh*-phrase is marked for singularity, the infinitival modality is equivalent to *should*.

(256) Hafdis knows which linguist to talk to at the party.
$\approx$ Hafdis knows which linguist she should talk to at the party.

Singular *wh*-phrases have the property that they presuppose the existence of a unique answer (cf. 257).

(257) a. Hafdis knows which linguist Penna likes.
(Presupposition: There is only one linguist that Penna likes.)

b. Hafdis knows which linguist she can talk to at the party.
(Presupposition: There is only one linguist that Hafdis can talk to at the party.)

Now let us consider an infinitival question with a singular *which* phrase.

(258) Hafdis knows which linguist to talk to at the party.
$[\![which\ linguist\ to\ talk\ to\ at\ the\ party]\!] =$
$\{\Diamond_{D,\rightarrow}(talk(hafdis,m)), \Diamond_{D,\rightarrow}(talk(hafdis,h)),$
$\Diamond_{D,\rightarrow}(talk(hafdis,p))\}$
(assumption: Magnus, Herb, and Penna are the only linguists in the domain)

Because of the presupposition of the *which* phrase, only one of the propositions in the question denotation of (258) can be true. Let us assume that the unique true proposition in the denotation of (258) is $\Diamond_{D,\rightarrow}(talk(hafdis,m))$. $\Diamond_{D,\rightarrow}(talk(hafdis,m))$ tells us that for Hafdis to talk to Magnus is a way of achieving the goals. From the presupposition of the *which*-phrase, we know that this is the only way of achieving the goals by talking to a linguist. But these two facts put together do not entail $\Box_D(talk(hafdis,m))$.

To go from $\Diamond_{D,\rightarrow}(talk(hafdis,m))$ to $\Box_D(talk(hafdis,m))$, we need a proposition that is stronger than what is given to us by the presupposition of the *which*-phrase. This proposition is shown in (259).

(259) $\forall p'[p' \neq p \Longrightarrow \neg\Diamond_{D,\rightarrow}(p')]$
(No proposition that is distinct from *p* is a way of achieving the goals)
$\forall p[[\Diamond_{D,\rightarrow}(p) \wedge \forall p'[p' \neq p \Longrightarrow \neg\Diamond_{D,\rightarrow}(p')]] \Longrightarrow \Box_D(p)]$
(If *p* is a way of achieving the goals, and no proposition that is distinct from *p* is a ways of achieving the goals, it follows that *p* is deontically necessary.)

$\Diamond_{D,\rightarrow}(talk(hafdis,m))$ being the only way of achieving the goals by talking to a linguist does not exclude there being ways of achieving the goals by talking to, for example, an archaeologist. On the other hand, (258) seems to presuppose that there is only way of achieving the goals. If the goals can be achieved by talking to a certain linguist and can also be met by talking to a certain archaeologist, (258) seems inappropriate. We have seen that the presupposition of the *which* phrase is on its own insufficient to explain the appropriateness of the *should* paraphrase. So where does this presupposition come from?

We trace this presupposition to an alternate source: the relationship of causation that holds between the propositional complement of $\Diamond_{D,\rightarrow}$ and the goals and the way in which the contextual restriction *C* is generated. Let us assume in the context of (258) that there also exists a way of meeting the goals which does not involve talking to a linguist i.e. the following propositions are true: $\Diamond_{D,\rightarrow}(talk(hafdis,m))$, $\Diamond_{D,\rightarrow}(talk(hafdis,k))$ (*k* is an archaeologist, not a linguist). What I will show now is that $\Diamond_{D,\rightarrow}(talk(hafdis,m))$ and $\Diamond_{D,\rightarrow}(talk(hafdis,k))$ cannot be true simultaneously without there being a 'presupposition' failure.

(260) Background assumption: Hafdis's talking to the archaeologist *k* will lead the achievement of the goals.
$$\Diamond_{D,\rightarrow}(talk(hafdis,m)) \Longleftrightarrow$$
$$\exists w'[talk(hafdis,m,w') \wedge w' \in Goal(w)] \wedge$$
$$\forall w'[[w' \in C \wedge talk(hafdis,m,w')] \rightarrow_c w' \in Goal(w)]$$

By the positive constraint on *C*, *C* must contain a world from each cell of the partition corresponding to the question *who to talk to at the party*. Due to the presupposition of *which*, $\Diamond_{D,\rightarrow}(talk(hafdis,m))$ will be true in only one cell and for our purposes only this one cell will matter. Now by the charity constraint on C, the worlds picked from the relevant cell of the partition will be worlds where everything that is conducive to the satisfaction of the goals is done. Hafdis's talking to the archaeologist k is something conducive to the satisfaction of the goals and so all the worlds picked from the relevant cell will have Hafdis talking to the archaeologist k.

But now there will be no causality relationship (represented here as $\rightarrow_c$) between Hafdis's talking to the linguist Magnus and the achievement of the goals. The goals will be satisfied independently by Hafdis's talking to the archaeologist k. So $\Diamond_{D,\rightarrow}(talk(hafdis,m))$ and $\Diamond_{D,\rightarrow}(talk(hafdis,k))$ cannot simultaneously be true in the context of (258). This is why it is infelicitous to use (258) if there are other ways of achieving the goals.

The presupposition that there is no non-linguist way of achieving the goals together with the fact that $\Diamond_{D,\rightarrow}(talk(hafdis,m))$ is the only linguist way of achieving the goals entails $\Box_D(talk(hafdis,m))$ (cf. 259). Thus we see why *should* paraphrases are appropriate for infinitival questions with singular *wh*-phrases.

3.4.2. *Contextually licensed entailments*

In certain contexts, *should* paraphrases are available with non-singular *wh*-phrases also. These contexts are cases where the true question denotation contains only element. Consider (261).

(261) Hafdis knows who to talk to at the party.
Context: the only way to become popular is by talking to Magnus, Herb, and Penna. Assume for simplicity that they are the only people in the domain.
$[\![\textit{who to talk to at the party}]\!] =$
$\{\Diamond_{D,\rightarrow}(talk(hafdis,m)), \Diamond_{D,\rightarrow}(talk(hafdis,h)),$
$\Diamond_{D,\rightarrow}(talk(hafdis,p)), \Diamond_{D,\rightarrow}(talk(hafdis,m+p)),$
$\Diamond_{D,\rightarrow}(talk(hafdis,m+h)), \Diamond_{D,\rightarrow}(talk(hafdis,h+p)),$
$\Diamond_{D,\rightarrow}(talk(hafdis,m+h+p))\}$

The only true element in the question denotation is $\Diamond_{D,\rightarrow}(talk(hafdis,m+h+p))$.[62] Because the restriction of *who* picks out all the people who could be talked to at the party, it turns out to be the case that Hafdis talking to Magnus, Herb, and Penna is the only way of achieving the goals. Therefore it follows (cf. 259) that $\Box_D(talk(hafdis,m+h+p)))$ is true. Consequently a *should* paraphrase is available.

In (261), a *should* paraphrase was available because the restriction of *who* picked out all the people at the party. This is not the case in (262). A *should* paraphrase, nevertheless, seems available.

(262) Hafdis knows which linguists to talk to at the party.
Context: the only way to become popular is by talking to the linguists Magnus, Herb, and Penna. Assume for simplicity that they are the only people in the domain.
$[\![which\ linguists\ to\ talk\ to\ at\ the\ party]\!] =$
$\{\Diamond_{D,\rightarrow}(talk(hafdis,m)), \Diamond_{D,\rightarrow}(talk(hafdis,h)),$
$\Diamond_{D,\rightarrow}(talk(hafdis,p)), \Diamond_{D,\rightarrow}(talk(hafdis,m+p)),$
$\Diamond_{D,\rightarrow}(talk(hafdis,m+h)), \Diamond_{D,\rightarrow}(talk(hafdis,h+p)),$
$\Diamond_{D,\rightarrow}(talk(hafdis,m+h+p))\}$

The only true element in the question denotation is $\Diamond_{D,\rightarrow}(talk(hafdis,m+h+p))$. We have seen in the discussion of infinitival questions with singular *which* phrases that there is a presupposition that there is only one way to achieve the goals and that this way involves a member of the restriction of the *which* phrase. Using this result in (262), we can go from Hafdis talking to Magnus, Herb, and Penna being the only linguist-way of achieving the goals to it being the only way of achieving the goals. It follows (cf. 259) that $\Box_D(talk(hafdis,m+h+p)))$ is true. A *should* paraphrase is therefore available.

3.4.3. *Neither* should *nor* could

We have seen that when the second conjunct in the definition of $\Diamond_{D,\rightarrow}$ is trivially true, $\Diamond_{D,\rightarrow}$ reduces to $\Diamond_C$. When $\Diamond_{D,\rightarrow}(p)$ is the only true proposition of its form (i.e. for all $p' \neq p$, $\Diamond_{D,\rightarrow}(p')$ is false), it entails $\Box_D$. When neither of these conditions hold, a true answer to an infinitival question involves neither a *could* nor a *should*. This is the case for the example in (263).

(263) Hafdis knows who to talk to at the party.
Context: the only way to become popular is by talking to Magnus, Herb, and Penna or by talking to Daniel, Stefan-Árni, and Baldur
true members of $[\![who\ to\ talk\ to\ at\ the\ party]\!] =$
$\{\Diamond_{D,\rightarrow}(talk(hafdis,m+h+p)), \Diamond_{D,\rightarrow}(talk(hafdis,d+s+b))\}$

Somewhat surprisingly though, if one tries to paraphrase (263) in the context indicated, a *should* paraphrase seems more appropriate than a *could* paraphrase. The reason a *should* paraphrase sounds appropriate is that modal questions with *should* permit a reading which is equivalent to (263).

Let us make a brief digression into the readings available with *should* questions. Consider the modal question in (264).

(264) Didda knows [who she should talk to at the party]
$[\![\textit{who she should talk to at the party}]\!] =$
$$\lambda w' \lambda p \exists x [R(x,w') \wedge p = \lambda w \Box_D(talk(didda,x,w))]$$
Context 1: Didda should talk to Olafur at the party.
True answers in Context 1: $\{\Box_D(talk(didda,o))\}$
Context 2: Didda should talk to Olafur or Penna at the party.
True answers in Context 2: $\{\}$

There is a sense in which in context 2, the embedded question in (264) does not have an appropriate answer. There is no x such that Didda should talk to x. But there is a *should* proposition which seems like a reasonable answer to the question *who she should talk to at the party* (cf. 265).

(265) Didda should talk to Olafur or Penna at the party.
$\Box_D(talk(didda,o) \vee talk(didda,p))$

How can we compositionally generate question denotation of the form in (265)? One way to do this is by assuming that *who* can leave behind a property trace of type $< e,t >$ instead of the individual trace of type e that it leaves behind in (264).

(266) $[\![\textit{who she should talk to at the party}]\!] =$
$\lambda w' \lambda p \exists \mathscr{P} [R(\mathscr{P},w') \wedge p = \lambda w \Box_D(\exists x [\mathscr{P}(x) \wedge talk(didda,x,w)])]$

Consider possible values of $\mathscr{P}$. Suppose $\mathscr{P}$ picks out only Olafur. Then such a $\mathscr{P}$ will not appear among the true answers because it is not the case that in all the relevant worlds, Didda talks to Olafur. The $\mathscr{P}$ that picks out Olafur and Penna will be a true answer. In addition, any $\mathscr{P}$ that picks out a set that contains Olafur and Penna will also be a true answer. So we have to have a way to pick out the 'most informative' $\mathscr{P}$.

Knowing that $\Box_D(\exists x[\mathscr{P}(x) \wedge talk(didda,x,w)])$ for the $\mathscr{P}$ that picks out Olafur and Penna gives us a weakly exhaustive reading. This weakly exhaustive answer looks like the following:

(267) $\Box_D(p_1 \vee \ldots \vee p_n)$

It seems that it is possible to relax the weak exhaustivity requirement in certain cases. In the cases where the weak exhaustivity requirement can be relaxed, it is sufficient for Hafdis to know (268a), or more precisely a true member of the question denotation in (268b).

(268) a. $\Box_D(\ldots \vee p_i \vee \ldots)$

b. Non-exhaustive answers are picked out by the following denotation:
Non-exhaustive $[\![\textit{who to talk to at the party}]\!] =$
$\lambda w' \lambda p \exists \mathscr{P}[R(\mathscr{P}, w') \wedge$
$p = \lambda w \exists \mathscr{Q}[\Box_D(\exists x[(\mathscr{P}(x) \vee \mathscr{Q}(x)) \wedge talk(PRO, x, w)])$
$\wedge \neg \Box_D(\exists x[\mathscr{Q}(x) \wedge talk(PRO, x, w)])]]$

(269) provides an illustration of the relaxation of exhaustivity that is formalized in (268b).

(269) Didda knows who she should talk to at the party.
Context: It is for the satisfaction of the goals that she should talk to Magnus, Herb and Penna or to Daniel, Siggi, and Baldur. The goals will be satisfied either way.

(269) is not judged false if Didda knows that talking to Magnus, Herb, and Penna will lead to satisfaction of the goals. It is not necessary for Didda to also know that talking to Daniel, Siggi, and Baldur will lead to satisfaction of the goals. If the answer had to be exhaustive, (269) would be judged false in the context provided. [63]

The non-exhaustive readings available with *should* questions are equivalent to the readings available with (263). This is why despite the absence of an equivalence with a *should* proposition, a paraphrase by a *should* question seems appropriate. [64]

3.5. Interaction with negation

The presence of negation in the embedded infinitival clause brings out the *should* reading.

(270) a. Hafdis knows [where not to get gas].

b. Hafdis knows [where to get gas].

The goals in (270) involve not getting gas from a place that is inappropriate - the place may be selling low-quality gas, overcharging, supporting polictically incorrect causes etc. Unlike (270b), a *could* reading does not seem to be available for (270a). We showed that a *could* reading became available for (270b) because there was trivial satisfaction of the goals. The goals for (270b) being the acquisition of gas, any place where one got gas would lead to satisfaction of the goals. Thus the second conjunct in the definition of $\Diamond_{D,\rightarrow}$, repeated in (271), is trivially satisfied and $\Diamond_{D,\rightarrow}$ reduces to circumstantial possibility $\Diamond_C$.

(271) $[\![\Diamond_{D,\rightarrow}(p)]\!]^w = 1 \iff \exists w'[w' \in Goal(w) \wedge p(w')] \wedge$
$\forall w'[[w' \in C \wedge p(w')] \rightarrow_c w' \in Goal(w)]$

Special case of $\Diamond_{D,\rightarrow}$
If $\forall w'[[w' \in C \wedge p(w')] \rightarrow_c w' \in Goal(w)]$ is trivially true, $\Diamond_{D,\rightarrow}$ becomes equivalent to circumstantial possibility:
$[\![\Diamond_{D,\rightarrow}(p)]\!]^w = 1 \iff \exists w'[w' \in Rel(w) \wedge p(w')] \iff \Diamond_C(p)$

The *could* reading is not available in the presence of negation because the goals are not trivially satisfied by not getting gas at a particular location. (270a) is considered in greater detail below.

The question denotation of the infinitival question *where not to get gas* is shown in (272). Note that in (272) the modality takes scope over negation. This follows from our syntactic assumption that the source of the modality in infinitival questions is the infinitival complementizer C^o. The complementizer takes scope over the negation yielding the question denotation in (272).

(272) Assume there are only three gas stations *a*, *b*, and *c*. *c* is the only acceptable place given the goals.
$[\![$*where not to get gas*$]\!] =$
$\{\Diamond_{D,\rightarrow}(\neg getGas(h,a)), \Diamond_{D,\rightarrow}(\neg getGas(h,b)), \Diamond_{D,\rightarrow}(\neg getGas(h,c)),$
$\Diamond_{D,\rightarrow}(\neg getGas(h,a+b)), \Diamond_{D,\rightarrow}(\neg getGas(h,b+c)),$
$\Diamond_{D,\rightarrow}(\neg getGas(h,a+c)), \Diamond_{D,\rightarrow}(\neg getGas(h,a+b+c))\}$

The Groenendijk & Stokhof style partition corresponding to this question is in (273).

(273) a. Goals not met:

i. $getGas(h,a), \neg getGas(h,b), \neg getGas(h,c)$
ii. $\neg getGas(h,a), getGas(h,b), \neg getGas(h,c)$
iii. $getGas(h,a), getGas(h,b), \neg getGas(h,c)$

iv. $\neg getGas(h,a), getGas(h,b), getGas(h,c)$

v. $getGas(h,a), \neg getGas(h,b), getGas(h,c)$

vi. $getGas(h,a), getGas(h,b), getGas(h,c)$

b. Goals met:

i. $\neg getGas(h,a), \neg getGas(h,b), \neg getGas(h,c)$

ii. $\neg getGas(h,a), \neg getGas(h,b), getGas(h,c)$

A *could* paraphrase is not suitable for (270a) since not getting gas at say, gas station *a*, does not trivially satisfy the goals. The goals associated with (270a) involve not getting gas from a place that is inappropriate. So if Hafdis does not get gas from gas station *a*, she could still end up getting from gas station *b*, which would also not be appropriate. Let us go through a derivation to see why $\Diamond_{D,\rightarrow}(\neg getGas(h,a))$ and $\Diamond_{D,\rightarrow}(\neg getGas(h,b))$ are not true in the context in (272).

(274) A necessary condition for $\Diamond_{D,\rightarrow}(\neg getGas(h,a))$ to be true is that in all worlds in C where Hafdis does not get gas from *a*, the goals should be met.
C contains worlds, where Hafdis does not get gas from *a* but gets gas from *b* (e.g. a world from cell c of the partition in 273). The goals are not met in such worlds thereby falsifying $\Diamond_{D,\rightarrow}(\neg getGas(h,a))$

A *should* paraphrase is appropriate for the following reason: The only true members of the question denotation in (272) are $\Diamond_{D,\rightarrow}(\neg getGas(h,a+b))$ and $\Diamond_{D,\rightarrow}(\neg getGas(h,a+b+c))$. The latter is ruled out as being 'less informative'. Consequently we have a unique answer.[65] We saw in the discussion of *should* paraphrases that the existence of unique answers entailed a *should* paraphrase. Hence the availability of a *should* paraphrase is to be expected.

It is puzzling that a *could* reading is never available with a negated infinitival question. It should be possible to construct an example where the goals are satisfied by the mere non-performance of an event. The examples that come to my mind are very artificial. cf. *Hafdis knows where to not get gas*. Imagine that *not getting gas* is an activity and the goal is merely to have taken part in *not getting gas*. Then a *could* reading is available. However, in this example the negation seems to be interpreted as part of the event itself. We speculate that the 'satisfaction by non-performance' readings are hard to access because of their extremely weak truth conditions.

3.6. Infinitival *how many*-N/*how*-Adv/Adj/*how much* questions

Infinitival questions where the extracted element is a degree phrase seem to only permit the *should* reading.[66]

(275) a. The NATO spokesman knows [how much to say about the bombing of civilian facilities].

b. Penna knows [how many people to invite to the party].

c. Olafur knows [how detailed to make his presentation].

A *could* reading is unavailable because there is no trivial satisfaction of the goals. The goal is not (merely) to say a certain amount about the bombing of civilian facilities, or to invite a certain number of people etc.

In order to apply the semantics developed for infinitival questions to infinitival degree questions, we need to decide what kind of trace is left behind in a degree extraction. Two options will be entertained: (i) the trace of degree extraction is interpreted as a variable over degrees (Sect. 3.6.1), and (ii) the trace of degree extraction is interpreted as a variable over sets of degrees (Sect. 3.6.2). The choice between these two options has empirical consequences which will be discussed.

3.6.1. *Infinitival degree questions: variables over degrees*

Let us consider what the question denotation of the infinitival question in (275a) consists of if we assume that the trace left behind by a degree extraction is interpreted as variable over degrees[67].

(276) $$\{\Diamond_{D,\rightarrow}(\exists x[invite(penna,x) \wedge Card(Atom(x)) = 1]), \\ \Diamond_{D,\rightarrow}(\exists x[invite(penna,x) \wedge Card(Atom(x)) = 2]) \\ \ldots\}$$

(*Atom* is a function that maps a potentially plural element into the set that consists of its constituent atoms. *Card* maps sets to their cardinality.)

Let us assume that one of the propositions in the question of (275a) is true, say $\Diamond_{D,\rightarrow}(\exists x[invite(penna,x) \wedge card(Atom(x)) = k])$. The partition corresponding to the question in (275a) is shown below in (277).

(277) Penna invites exactly 1 person.
Penna invites exactly 2 people.
...

If $\Diamond_{D,\rightarrow}(\exists x[invite(penna,x) \wedge card(Atom(x)) = k])$ is true, it follows from the semantics of $\Diamond_{D,\rightarrow}$ that must also be the case that for all $n \geq k$, it is also true that $\Diamond_{D,\rightarrow}(\exists x[invite(penna,x) \wedge card(Atom(x)) = n])$. This result seems too strong. It may be correct in some cases, where all that is required for the satisfaction of goals is that there be at least a certain number of people at the party. However, in general, it is possible that it is necessary for the satisfaction of the goals that exactly *k* people be invited. There seems to be no way to represent the answer 'exactly *k*' given the question denotation in (276).

The problem seems to lie in the way we interpret the degree variable. So far we have been assuming that the degree variable in the question denotation has the downward entailment property i.e. if Penna invites *n* people to the party, for all $m \leq n$, she invites *m* people to the party. If we assume maximalization in each question denotation, this problem does not arise (cf. 278). (cf. Rullmann 1995 for a detailed discussion of the role played by maximalization in *wh*-constructions.)

(278) $\{\Diamond_{D,\rightarrow}(Card(Atom(\iota(\lambda x.invite(penna,x)))) = 1),$
$\Diamond_{D,\rightarrow}(Card(Atom(\iota(\lambda x.invite(penna,x)))) = 2),$
$\ldots\}$

Assume $\Diamond_{D,\rightarrow}(Card(Atom(\iota(\lambda x.invite(penna,x)))) = k)$ is a true proposition in (278). The partition corresponding to the question denotation in (278) will be the same as the partition corresponding to the question denotation in (276). However, now there will be only one cell in the partition with which $\Diamond_{D,\rightarrow}(Card(Atom(\iota(\lambda x.invite(penna,x)))) = k)$ will be compatible. So the problem of undesirable entailments that arose with the question denotation in (276) will not arise.

Building maximalization into the individual question denotations also helps us to analyze cases where more than one answer is appropriate.

(279) Hafdis knows how many points to score.
Context: A game where the goal is to score 10 points or 20 points. Hafdis has 7 points. So if she makes 3 points or 13 points, she will win. If she makes more or less, she will lose.

For (279) to be judged true in the context indicated, Hafdis has to know that scoring 3 points will lead to achievement of the goals or that scoring 13 points will lead to achievement of the goals. She does not have to know both. The true members of the question denotation are shown in (280).

(280) $\{\Diamond_{D,\rightarrow}(\iota(\lambda x.score(hafdis,x)) = 3),$
$\Diamond_{D,\rightarrow}(\iota(\lambda x.score(hafdis,x)) = 13)\}$

Hafdis's knowing any of the true members of the question denotation, as shown in (280), suffices for (279) to be judged true.

3.6.2. *Infinitival degree questions: variables over sets of degrees*

Now let us consider what the question denotation of the infinitival question in (275a) consists of if we assume that the trace left behind by a degree extraction is interpreted as variable over sets of degrees[68]. In particular, the variable left behind is interpreted as a variable over intervals. Intervals are sets of degrees that have the property in (281).

(281) $\forall I \forall d \forall d'[d \in I \wedge d' \in I \rightarrow$
$\forall d''[[[d \leq d'' \leq d'] \vee [d' \leq d'' \leq d]] \rightarrow d'' \in I]]$

The interval-speak that we will be employing in this discussion can be related to degree-speak (which we are more used to) in the following fashion:

(282) $tall(x,D) \Leftrightarrow \forall d[d \in D \rightarrow tall'(x,d)]$
x is D-tall: x is tall to every degree that lies in the interval D.

Note that we are talking about intervals is not committing us to maximality. So if John is tall to the interval (0,6ft.), he is also tall to the interval (0,5ft.).

The question denotation of (275a) under the suggested interpreted of the degree trace is shown in (283). This is the most general representation. In most cases, it might be the case that m is set to zero.

(283) $\{\Diamond_{D,\rightarrow}(\exists x[invite(penna,x) \wedge m \leq card(Atom(x)) \leq n]), \ldots, \ldots\}$
(for different values of m, n)

Let us assume that one of the propositions in the question denotation in (283) is true, say $\{\Diamond_{D,\rightarrow}(\exists x[invite(penna,x) \wedge m \leq card(Atom(x)) \leq n])$. The partition corresponding to the question in (275a) is shown below in (284).

(284) Penna invites exactly 1 person. (call this class of worlds 'cell w_1')
Penna invites exactly 2 people. (call this class of worlds 'cell w_2')
...
Penna invites exactly *n* people. (call this class of worlds 'cell w_n')
...

Note that the change in the question denotation does cause a change in what parts of the partition we need to examine. While checking for the truth of $\Diamond_{D,\rightarrow}(\exists x[invite(penna,x) \wedge m \leq card(Atom(x)) \leq n])$, we need to only look at the worlds in cells w_m through w_n. w_m through w_n are the only cells where the antecedent of the second conjunct in the definition of $\Diamond_{D,\rightarrow}$ (repeated in 285) is true.

(285) $\Diamond_{D,\rightarrow}(p) \Longleftrightarrow$

$$\exists w'[w' \in Goal(w) \wedge p(w')] \wedge$$
$$\forall w'[[w' \in C \wedge p(w')] \rightarrow_c w' \in Goal(w)]$$

In all other worlds, $\exists x[invite(penna,x) \wedge m \leq card(Atom(x)) \leq n]$ will be false and the consequent will be true. By the conditions on C, C has to contain representative worlds from each cell in the partition. Only the representative worlds from cells w_m through w_n matter. For $\Diamond_{D,\rightarrow}(\exists x[invite(penna,x) \wedge m \leq card(Atom(x)) \leq n])$ to be true, it must be the case that in all of these worlds the goals are met. This seems reasonable. If the goals are met by inviting exactly 5 people, the only true proposition in the question denotation will be the one corresponding to the degenerate interval $[5 \leq x \leq 5]$. If as in (279), scoring 3 points or scoring 13 points both lead to goal satisfaction, the question denotation will contain two true propositions. One of the propositions will correspond to the degenerate interval $[3 \leq x \leq 3]$ and the other to the degenerate interval $[13 \leq x \leq 13]$.

3.6.3. *Comparison between the two approaches and other issues*

If we interpret the variable as a degree, as in Sect. 3.6.1, we need to introduce maximalization in the semantics to get the right results. This is conceptually troubling in two ways. Firstly, at a technical level - the maximalization that we need takes place below the modality. In the cases of maximalization discussed by Rullmann (1995), maximalization applies at the topmost level. Second, and more troubling, is the appeal to maximalization in degree questions alone. An important part of the analysis of infinitival questions proposed here is the absence of (even weak) exhaustivity. Exhaustivity and Maximalization have been analyzed by Rullmann (1995) as reflexes of the same mechanism. Appealing to one when explicitly disavowing the other is problematic. Adopting Schwarzschild (1998)'s independently motivated proposal regarding the interpretation of traces of degree extraction gives us the desired results without necessitating appeals to maximalization.

should paraphrases are available for infinitival degree questions for the reasons discussed in Sect. 3.4.2 and Sect. 3.4.3. If it is presupposed that there is only one true answer, a *should* paraphrase become available following the discussion in Sect. 3.4.2. If there is more than one true answer, a *should* paraphrase becomes available following the discussion in Sect. 3.4.3. *could* paraphrases are not available because the truth of the proposition does not trivially lead to goal satisfaction. Suppose inviting 5 people leads to satisfaction of the goals. Then since a world in which 5 people are invited is also a world in which 4 people are invited, it is true that a world in which four people are invited is a world in which the goals are satisfied. But inviting 4 people does not by itself lead to goal satisfaction. The second conjunct in the definition of the modality still does some work for us blocking a pure possibility interpretation.

4. Variable modal force in infinitival relatives

Hackl & Nissenbaum (1999) propose the generalization in (286).

(286) Determiner Modal Generalization: Strong determiners (and strong interpretations of weak determiners) induce a *should*-reading in *for*-infinitival relative clauses while weak determiners allow both *should* and *could* readings.

Here we will show how the semantics provided for the covert modality in infinitival questions can be used to explain the Determiner Modal Generalization. [69] We will see how the same ingredients play a role in determining the modal force of the covert modality in infinitival questions and infinitival relatives.

It will be assumed throughout this discussion that the covert modality ($\Diamond_{D,\rightarrow}$) that was assumed to appear in the C^0 of infinitival questions also appears in the the C^0 of infinitival relatives (cf. Chap. 2). It is therefore to be expected that infinitival relatives, like infinitival questions, show variable modal force effects.

Our explanation of the Determiner Modal Generalization will derive the *should* readings induced by strong quantifiers and the strong readings of weak quantifiers and the the *should* readings available with weak readings of weak quantifiers from distinct sources. The *should* readings found with strong quantifiers and the strong readings of weak quantifiers have a 'structural' source while the *should* readings available with weak readings of weak quantifiers have a 'contextual' source. Recall that with infinitival questions also, some *should* readings had 'structural' sources (*which*-phrase) and others had 'contextual' sources (*should* readings of *where to get gas*).

4.1. *should* readings: the core case

We treat the *should* readings found with strong quantifiers and with strong readings of weak quantifiers as the core case of the Determiner Modal Generalization. Let us start with the case of *the* with a singular head noun.

4.1.1. the *with singular N*

I assume that the trace inside the relative clause is that of a singular individual variable. Also for simplicity, I will assume that the head NP is interpreted relative clause-externally. Nothing depends upon this assumption.

(287) the book [Op_i [PRO to read t_i]]
$\iota\lambda x[book(x) \wedge \Diamond_{D,\rightarrow}(read(PRO,x))]$

Two contexts in which the NP in (287) can be evaluated are considered below.

Context 1:
There are several books, $a,\ldots,e$, with the property $\lambda x.\Diamond_{D,\rightarrow}(read(PRO,x))$. The set to which ι applies is not a singleton set, nor does it contain a maxima. Therefore the result of applying ι to this set is undefined. Context 1 is not suitable.

Context 2:
There is a unique book, call it b with the property $\lambda x.\Diamond_{D,\rightarrow}(read(PRO,x))$. The set to which ι applies is a singleton set. Applying ι to this set returns us b.

Now recall that the uniqueness imposed by *which*-phrases in infinitival questions yielded a *should* paraphrase.

(288) Hafdis knows [which book to read]
$\approx$ Hafdis knows which book she/one should read.

The particular semantics of $\Diamond_{D,\rightarrow}$ allows us to go from the presupposition of the *which*-phrase (= There is a unique book s.t. reading that book leads to goal satisfaction) to 'There is a unique x s.t. reading that book x leads to goal satisfaction' i.e. reading book x is the only way to satisfy the goals. [70] Therefore a *should* paraphrase is suitable.

The uniqueness contributed by the semantics of *the* in an infinitival relative plays the same role as the uniqueness imposed by *which*-phrases in infinitival questions and consequently yields a *should* paraphrase too.[71]

4.1.2. the *with plural N*

I assume that the trace inside the relative clause is that of a plural individual variable.

(289) the books [Op$_i$ [PRO to read t$_i$]]
$\iota\lambda x[books(x) \wedge \Diamond_{D,\rightarrow}(read(PRO,x))]$

As with questions, superfluous supersets have to be excluded.

Context 1: Reading books a and b will lead to goal satisfaction. Reading books c and d will also lead to goal satisfaction. i.e. $\Diamond_{D,\rightarrow}(read(PRO,a+b))$ and $\Diamond_{D,\rightarrow}(read(PRO,c+d))$ are true. The set to which ι applies contains $a+b$ and $c+d$. Due to exclusion of superfluous supersets, $a+b+c+d$ is not part of the set to which ι applies. Consequently, there is no maxima and the result of applying ι to this set is undefined.

Context 2: Reading books a and b will lead to goal satisfaction. No other books will do. The set to which ι applies is a singleton set containing only $a+b$. (Note that $\Diamond_{D,\rightarrow}(read(PRO,a))$, $\Diamond_{D,\rightarrow}(read(PRO,b))$ are not true). ι-application returns $a+b$. Again due to the uniqueness contributed by *the*, we can go from $\Diamond_{D,\rightarrow}(read(PRO,a+b))$ to $\Box_D(read(PRO,a+b))$ (= the *should* reading).

4.1.3. *Strong quantifiers and strong readings of weak quantifiers*

We assume that all strong quantifiers and all strong readings of weak quantifiers involve covert partitives. In other words, strong quantifiers are always partitives and weak quantifiers can optionally be interpreted as partitives. The strong reading of weak quantifiers involves weak quantifiers interpreted as partitives.

(290) every N = every of the N
most N = most of the N
some$_{strong}$ N = some of the N

The covert partitive approach allows us to reduce all strong quantifiers and strong readings of weak quantifiers to the core case with *the* (cf. 287 and 289).[72]

To see why we need to treat quantifiers like *every* as a covert partitive, let us see what happens if we do not.

(291) Every man [Op$_i$ (for Miguel) to play against t$_i$] is in the next room.
$\forall x[[man(x) \wedge \Diamond_{D,\rightarrow}(play(m,x))] \Longrightarrow NextRoom(x)]$

(291) says that every person *x* such that playing against *x* would lead to satisfaction of goals is in the next room. This is not the reading we want for (291). Our intuition for (291) is that for goal satisfaction, Miguel has to play against the entire bunch of men in the next room. If *every N* is treated as *every of the Ns*, we get the desired results.

(292) Every man [Op$_i$ (for Miguel) to play against t$_i$] is in the next room.
= Every of the men [Op$_i$ (for Miguel) to play against t$_i$] is in the next room.
$\forall x[Atom(x, \iota(\lambda y[men(y) \wedge \Diamond_{D,\rightarrow}(play(m,y))])) \Longrightarrow NextRoom(x)]$

Consider now the contrast in (293).

(293) (from Hackl & Nissenbaum 1999) (uttered in a society where polygamy is not permitted)

a. # Norman's mother saw every/most/two of the women for him to marry at the party
(≈ At the party, Norman's mother saw every (etc.) women that he *should* marry.)

b. Norman's mother saw many/few/some/two women for him to marry at the party
(≈ At the party, Norman's mother saw many (etc.) women that he *could* marry.)

(293a) is felt to be odd because it is felt that Norman is supposed to marry a plurality of women, which is unusual in a monogamous society. (293b) lacks this feeling of oddness because all that is said is that a particular set of women are appropriate for Norman to marry. We are able to explain this contrast within our system.

the women for him to marry is a subconstituent of all the object NPs in (293a). To avoid a presupposition failure, the predicate *women for him to marry* must have a maxima. There must be a maximal plural individual x s.t. x consists of women and $\Diamond_{D,\rightarrow}(marry(n,x))$ is true. $\Diamond_{D,\rightarrow}(marry(n,x))$ requires that Norman marry a plurality of women leading to oddness.

For well-formedness, (293b) does not require there to be a maxima. The quantifiers *many/few/some/two* quantify over the set of women x such that $\Diamond_{D,\rightarrow}(marry(n,x))$ is true. x does not have to be a plural individual and therefore Norman does not have to marry a plurality of women. Consequently, (293b) is not odd in the way (293a) is. [73]

4.2. Weak readings of weak quantifiers

As part of their Determiner Modal Generalization, Hackl & Nissenbaum (1999) note that weak readings of weak quantifiers permit both *could* and *should* readings.

(294) (ex. from Hackl & Nissenbaum 1999)
There are many problems (for us) to write about
... and we're glad to have all the choices. (*could*)
... and we're stressed out by the obligation. (*should*)

Within the system assumed here, the *could* reading is a $\Diamond_{D,\rightarrow}$. Assuming that the covert modality in (294) is $\Diamond_{D,\rightarrow}$ (and not $\Diamond_{D}$, as Hackl & Nissenbaum have it) does not seem to create any problems.[74]

The *should* reading arises if it follows from the context that the set of problems to write about has a maxima. The existence of a maxima is equivalent to uniqueness. We know from the discussion of infinitival *which* questions and infinitival relatives with strong quantifiers that the presence of uniqueness allows for *should* paraphrases.

4.3. A brief comparison with Hackl & Nissenbaum 1999

Hackl & Nissenbaum (1999) provide a detailed analysis of object infinitival relatives and an explanation of the Determiner Modal Generalization. We do not go into details of their explanation but will limit myself to noting some ways in which our proposal differs from theirs.

Hackl & Nissenbaum argue that the *should* and the *could* readings reflect different underlying structures. Infinitival relatives with *could* readings obligatorily involve a relative clause-internal interpretation of the head NP, while *should* readings permit both a relative clause-internal and a external interpretation of the head NP. They provide evidence for the the relative clause-internal interpretation of the head NP in the case of *could* readings through Condition C reconstruction effects (cf. 295).

(295) (exs. from Hackl & Nissenbaum 1999)

a. There are many papers by her_i students for Sabine_i to read this weekend.
(both *could*, *should*, no Condn. C violn. results from Reconstruction of head NP)

b. There are many papers by Sabine's_i students for her_i to read this weekend.
(*should*, # *could*, Reconstruction of head NP triggers Condn. C violation)

Within the system proposed here the difference between *could* and *should* interpretations does not translate into a structural difference. Hence, it is not surprising that the Condition C reconstruction data that Hackl & Nissenbaum bring up is not predicted by our analysis. We note, however, that the contrast in (295) seems to be somewhat and disappears altogether for our speakers in the minimally different (296).

(296) a. There are many movies by her_i friends for Mimi_i to watch at the Ritz this weekend.
(both *could*, *should*)

b. There are many movies by Mimi's friends for her_i to watch at the Ritz this weekend.
(both *could*, *should*)

The proper analysis of cases like (295) and the contrast between (295) and (296) is left for future work.

5. Conclusions

In this chapter we examined several questions concerning modality in infinitival questions. We began with the observation that all infinitival questions involve modality. The source of this modality was identified as the infinitival [+wh] C^0. We then contrasted the class of predicates that took infinitival question complements with the class of predicates that took finite question complements, noting that the first is a proper subset of the second. There are no predicates that exclusively subcategorize for infinitival questions but there are predicates that can take finite clause complements but not infinitival complements. We derived this asymmetry by noting that the meanings that can be expressed by infinitival questions - particular kinds of modal propositions - are a subset of the meanings that can be expressed by finite questions.

Next we examined the properties of the modality that appears in infinitival questions. This modality is in general deontic/bouletic. In a limited set of environments, the modality can also be circumstantial. However, it can never be epistemic modality. The force of the modality in an infinitival question seems to vary depending upon the environment. Infinitival questions with *where* and *how* seem to involve a *could* modality, while infinitival questions with *which* seem to involve a *should* modality. We proposed that the apparent variations in force and nature (deontic vs. circumstantial) notwithstanding, the modality in all infinitival questions is the same. The semantics of this modality (= $\Diamond_{D,\rightarrow}$) were formalized and we showed that the apparent variations of force and nature fall out as special cases of this modality. Finally we examined the infinitival modality found in object infinitival relatives and showed that the machinery used to explain variable force effects in infinitival questions extends to explain variable force effects in non-subject infinitival relatives.

Chapter 5
Ability modals and their actuality entailments

In the preceding chapters, we have investigated cases where we find modality in the interpretation but seemingly not in the input. Now we turn to the reverse case, a case where we seem to have an overt modal element in the input but there seems to be no modality in the output. The particular case that we will examine is that of ability modal though we will see that the phenomenon of disappearing modality is more general. We begin with the observation that the English ability modal *was able to* is ambiguous between two readings which can be paraphrased as 'managed to' and 'had the ability to'. Our analysis of this ambiguity is guided by the fact that these two readings are expressed by combining different aspectual markings with the ability modal in languages where the presence of genericity is marked morphologically. According to our analysis, the 'managed to' and the 'had the ability to' readings result from different aspectual specifications of an underlying predicate *ABLE*. We propose that *ABLE* has the semantics of an implicative verb like 'manage to' and that the 'had the ability to' reading is derived by combining *be able to* with a covert modal, a particular kind of *Gen* operator.

1. The ambiguity of *was able to*

(297) can be embedded in two quite different kinds of contexts as indicated in (298a, b).

(297) John was able to eat five apples in an hour.

(298) a. Yesterday, John was able to eat five apples in an hour. (past episodic)

b. In those days, John was able to eat five apples in an hour. (past generic)

(298a) implicates that John actually ate five apples in an hour. I will refer to this implication as the **actuality implication** (leaving open, at this point, the question of whether it is an implicature or an entailment) of (298a). If an ability attribution attributes an ability $\mathscr{P}$ to an individual x, the actuality implication corresponding to this ability attribution states that the ability is realized by x. This is shown in (299).

(299) a. Ability Attribution: $ABILITY\ (\mathscr{P})(x)$

b. Actuality Implication: $\mathscr{P}(x)$

(299) abstracts away from issues of tense and temporal modifiers. I assume that any temporal modifiers that modify the ability attribution in (299a) will be inherited by the actuality implication in (299b).

Cancellation of the actuality implication leads to a certain oddness.

(300) Last night, a masked assailant attacked me on my way home. I was able to wrestle him to the ground. #But I didn't do anything since I am a pacifist.

The two readings associated with *be able to* allow different interpretive possibilities for indefinite/bare plural subjects.

(301) A fireman was/Firemen were able to eat five apples.

a. Yesterday at the apple eating contest, a fireman was/firemen were able to eat five apples. (Past episodic, actuality implication, existentially interpreted subject)

b. In those days, a fireman were/firemen were able to eat five apples in an hour (Generic, no actuality implication, generically interpreted subject)

In its most natural interpretation, (301a) has a 'managed to' reading i.e. there is an actuality implication. The indefinite subject *a fireman* (or the bare plural subject *firemen*) receives an existential interpretation. On the other hand, (301b) has a 'had the ability to' reading. There is no actuality implication and the indefinite subject *a fireman* (or the bare plural subject *firemen*) can only be interpreted generically.

There is a link between the availability of a non-generic reading and the presence of an actuality implication. The 'managed to' reading (the actuality implication reading) of *be able to* is available in the simple past tense, but not in the simple present. In English, non-states in the simple past tense are ambiguous between a past generic and a past episodic reading, while non-states in the simple present tense only have a present generic reading. To investigate this link further, we now consider languages where generic readings are marked by aspectual morphology, typically imperfective aspect.

2. A correlation between aspect and actuality implications

In languages where imperfective aspect appears on generic sentences (sentences in the perfective aspect lack generic readings), we find that when the ability modal occurs with imperfective aspect, there is no actuality implication. However, when the ability modal occurs in the past perfective, there is an actuality implication. Further, in these languages, the actuality implication seems to be uncancelable so we will refer to it as the actuality entailment.

2.1. The crosslinguistic evidence

When the ability modal is in the imperfective, the assertion of the ability can be followed by a clause asserting that the ability was not actualized (cf. 302a). However, when the ability modal is in the past perfective, the modal assertion cannot be followed felicitously by a clause asserting that the ability was not actualized [75] (cf. 302b).

(302) Greek ability modals in the perfective:

a. *Borusa na sikoso afto to trapezi ala δen*
CAN.impfv.1s NA lift.non-pst-pfv.1s this the table but NEG
to sikosa
it lift.impfv
'(In those days), I could lift this table but I didn't lift it.'

b. *Boresa na tu miliso (# ala δen tu*
CAN.pst-pfv.1s NA him talk.non-pst-pfv.1s but NEG him
milisa)
talk.pst-pfv
'I was able to talk to John (but I did not talk to him).'

The presence of the actuality entailment when the ability modal is in the past perfective can also be seen by the contradictoriness of the examples in (303). The corresponding examples in the imperfective are not contradictory. The counterpart of (303a) with the ability modal in the imperfective means something like 'On three occasions, John could have killed Peter'. Similarly, the imperfective counterpart of (303b) means something like 'I could have killed myself'.

(303) Greek ability modals in the imperfective:

a. *# O Yanis borese na skotosi ton Petro 3 fores*
the John CAN.pst-pfv NA kill.non-pst-pfv the Peter 3 times
'John managed to kill Peter three times.'

b. *# Boresa na aftoktoniso*
CAN.pst-pfv NA kill-self
'I managed to kill myself.'

Like the Greek ability modal in the imperfective aspect, the Hindi-Urdu ability modal *sak* in the imperfective[76] lacks an actuality entailment. This can be seen by the fact that they can be followed by a clause asserting the non-actualization of the ability (cf. 304a). There is an actuality entailment when the ability modal is in the past-perfective. The assertion cannot be followed by a clause asserting the non-actualization of the ability (cf. 304b).

(304) Hindi-Urdu ability modals in the imperfective/habitual:

a. *Yusuf havaii-jahaaz uṛaa sak-taa hai/thaa (lekin vo*
Yusuf air-ship fly CAN-impfv be.Prs/be.Pst but he
havaii-jahaaz nahĩĩ uṛaa-taa hai/thaa
air-ship Neg fly-impfv be.Prs/Be.Pst
'Yusuf is/was able to fly airplanes but he doesn't/didn't fly airplanes.'

b. *Yusuf havaii-jahaaz uṛaa sak-aa (# lekin us-ne*
Yusuf air-ship fly CAN-Pfv but he-erg
havaii-jahaaz nahĩĩ uṛaa-yaa)
air-ship Neg fly-Pfv
'Yusuf could fly the airplane, but he didn't fly the airplane.'

The presence of an actuality entailment when the ability modal is in the past perfective is also demonstrated by the contradictoriness of (305a, and b).

(305) Hindi-Urdu ability modals in the perfective:

a. *# Yunus Yakub-kaa tiin baar khoon kar sak-aa*
Yunus Yakub-Gen 3 times murder do CAN-Pfv
'Yunus could murder Yakub three times/on three occasions.'

b. *# mE apne-aap-ko maar sak-aa*
I self-Acc kill can-Pfv
'I could kill myself.'

A very similar pattern has been noted for French in Hacquard (2005).

(306) French (from Hacquard (2005))

a. Imperfective:

Jean pouvait soulever cette table, mais il ne l'a pas soulevée.
Jean could-Impfv lift this table but he NE it-has Neg lifted
'Jean could lift this table, but he didn't lift it.'

b. Perfective:

Jean a pu soulever cette table, #mais il ne l'a pas soulevée.
Jean could-Pfv lift this table but he NE it-has Neg lifted
'Jean could lift this table, #but he didn't lift it.'

2.2. The general pattern

The pattern that emerges across the languages surveyed above where the perfective/ imperfective distinctions is marked on ability modals is shown in (307).

(307) a. Past (Pfv(**CAN**) [VP]) = *managed-to*
b. Past (Impfv(**CAN**) [VP]) = *had-ability-to*

The two readings associated with *be able to* in the past are realized in these languages by distinct forms. Across languages, imperfective aspect primarily makes two distinct semantic contributions - the semantics of an event in progress and the semantics of genericity. The semantic contribution relevant here is the semantics of genericity. This can be seen by the fact that the Hindi imperfective marker *-taa* only contributes the semantics of genericity, the 'event-in-progress' reading being marked by a specialized progressive marker.[77]

The pattern exemplified for French, Greek, and Hindi-Urdu also obtains in Bulgarian (Roumyana Pancheva p.c.), Catalan (Sergi Casals, Miguel-Angel Hernando-Cupido, Luis Lopez, Josep Quer, and Maria Isabel Oltra-Massuet p.c.). The facts in Albanian (Dalina Kallulli p.c.), Basque (Karlos Arregi-Urbina p.c.), Galician (Carmen Rio-Rey p.c.), Brazilian Portuguese (Luciana Storto p.c.), and Spanish (Olga Fernandez p.c.) are substantially similar though not identical to the pattern summarized above. They differ in the following manner: the languages accounted for in the main text have the following entailment 'John can-Pfv lift the table' entails 'John lifted the table'. However, in many languages e.g. Spanish, it seems possible to shift to a counterfactual meaning 'John could have lifted this table'. We do not have an analysis of why this additional reading is available but our speculation is that Hacquard (2005)'s observation about the ambiguity of the French *passé composé* between a preterit (which yields an actuality entailment) and a perfect (which does not) is likely to be relevant here.

3. The actuality implication and its relationship with ability

Does an ability modal with an actuality implication still have an ability component to its meaning? At first it seems that it does. However, looking at a wider array of facts suggests that an ability modal with an actuality implication does not have an ability component to its meaning.

3.1. An ability component in the actuality entailment?

Ability modal sentences with an actuality implication do not just mean that an event related to the embedded predicate took place. If they did, then the sentences in (308) would have been fine on the readings indicated in parentheses.

(308) a. *Yesterday, it was able to rain here. (Yesterday, it rained here)

b. #/* *baarish ho sak-ii* *(Hindi-Urdu)*
rain.f be CAN-Pfv.f
'# It was able to rain.'

c. # The mailman was able to be bitten by a dog yesterday. (The mailman was bitten by a dog yesterday)

Also, if all that was asserted was that an event related to the embedded predicate was actualized, the oddness of (309) would be puzzling.

(309) # A woman in Watertown was able to win 3 million dollars in the lottery yesterday.

A plausible explanation for the oddness of (308a, b) and (309) is that they do not constitute good ability attributions. This explanation presupposes that ability modal sentences with an actuality implication involve an ability attribution. However, the oddness of (309) vanishes if the context makes it clear that winning the lottery involves some kind of sustained (non-minimal) effort.

(310) After buying lottery tickets regularly for several years, a woman in Watertown was finally able to win 3 million dollars in the lottery yesterday.

The manner in which the oddness of (309) is alleviated in (310) suggests that ability modals with an actuality implication do not necessarily involve an ability attribution and that what was wrong with (309) was not that no ability was involved but that the context did not indicate that some effort went into the action. Once that was fixed, (309) (cf. 310) improved considerably (also see Sect. 3.3). It is harder to imagine a suitable situation involving sustained non-minimal effort when we are talking about meteorological events and this correlates with the seeming irredeemability of (308a, b).

3.2. Relationship with implicative verbs

The oddness of (309) and the manner in which this oddness is ameliorated is similar to the oddness of (311) involving the implicative verb *manage* and the manner of the amelioration of this oddness.

(311) a. # A woman in Watertown managed to win 3 million dollars in the lottery yesterday.
b. After buying lottery tickets compulsively for several years, a woman in Watertown finally managed to win 3 million dollars in the lottery yesterday.

We saw that the odd (309) and (311a) became acceptable, if the context indicates that some effort went into the action. How is this component of the meaning of *be able to* and *manage to* represented? We argue that this component of their meaning is represented as a conventional implicature. To this end, we introduce some tests developed by Karttunen & Peters (1979) for identifying conventional implicatures.

Karttunen & Peters (1979) argue that the meaning of an implicative verb like *manage* is best represented in terms of its assertion and its conventional implicature as in (312).

(312) John managed to sit through the Chinese opera.

a. Assertion: John sat through the Chinese opera.

b. Conventional Implicature: Sitting through a Chinese opera requires some effort for John.

They base this distinction upon the fact that the parts of the meaning of (312) indicated in (312a) and (312b) behave differently with respect to entailment patterns and presupposition projection. For example, (313a) entails (313b) but not (313c).

(313) a. I just discovered that John managed to sit through the Chinese opera.

b. I just discovered that John sat through the Chinese opera.

c. I just discovered that sitting through a Chinese opera required some effort for John.

The facts with *was able to* with an actuality implication seem similar.

(314) a. I just discovered that John was able to sit through the Chinese opera.

b. I just discovered that John sat through the Chinese opera.

c. I just discovered that sitting through a Chinese opera required some effort for John.

The pattern in (314) suggests that the part of the meaning of *was able to* that makes cases like (308a, b) and (309) odd is part of the conventional implicature of *was able to* and not part of its assertion.

3.3. More *was able to* without ability

(315b) is another case where we seem to have just an actuality entailment without an accompanying ability attribution.

(315) (from Thalberg 1969)

a. Yesterday, Brown hit three bulls-eyes in a row. Before he hit three bulls-eyes, he fired 600 rounds, without coming close to the bulls-eye; and his subsequent tries were equally wild.

b. Brown was able to hit three bulls-eyes in a row.

c. Brown had the ability to hit three bulls-eyes in a row.

From (315a), we can conclude (315b) but not (315c). Brown could have hit the target three times in a row by pure chance and he does not need to have had any ability for (315b) to be true.

If we accept that *was able to* can be used in the absence of any actual ability attribution, we have an argument that shows that treating actuality implications as implications of an associated ability attribution cannot be correct. As (310) and (315b) show, actuality implications are present even in the absence of an ability attribution to entail them. So the actuality implication cannot be the implication of an associated ability attribution. The actuality implication (and its associated conventional implicature) is all there is.

We can still ask the question of whether the actuality implication is cancelable or not and on the basis of (310) and (315b), we can say that it is not cancelable. Therefore, it is either part of the assertion (or entailed by the assertion) or part of the conventional implicature. The entailment pattern in (314) suggests that what we are calling the actuality implication constitutes the assertion of (314a). We will, however, continue to refer informally to the 'managed to' reading of *be able to* as the actuality entailment of the ability modal.

3.4. Conclusions concerning the meaning of ability modals

When an ability modal has an actuality entailment, there is no ability attribution. In fact the term 'actuality entailment' as defined in (299) is misleading because it suggests that there is an ability attribution that entails the actuality entailment. According to our analysis, in the cases where there is an actuality entailment, the actuality entailment and its associated conventional implicature is all we have. The associated conventional implicature requires that the relevant event involved some effort on the part of the subject. (308a, b) are unacceptable because in these cases, the subject - an expletive in (308a) and the inanimate *baarish* 'rain' in Hindi-Urdu - is not something that can make any kind of effort. The conventional implicature of the ability modal is not satisfied leading to unacceptability.

4. Compositional derivation

4.1. Existential readings with *be able to*

We have seen in (301) that bare plural/indefinite subjects of *was able to* can receive both a generic and an existential interpretation. The existential interpretation of a bare plural/indefinite subject of *was able to* was accompanied by an actuality entailment (cf. 301a). But is it, in general, necessary for an ability modal to have an actuality entailment, in order for its indefinite subject to be interpreted existentially? The following examples seem to be cases where there is no actuality entailment and yet the indefinite subject receives an existential interpretation.[78]

(316) a. Yesterday, at the apple eating contest, a fireman was able to eat fifty apples in an hour. I know because I had seen him drink an illegal performance enhancing potion. However, he never ate any apples at the contest because the judges caught him and barred him from the contest.

b. Last Tuesday, in Schenectady, a five year old girl was able to lift 500 kilograms. The scientists were able to detect her ability by measuring her muscle stress. She never actually lifted anything because she was straightjacketed throughout Tuesday to prevent her hurting herself. It is speculated that she had temporary acquired superhuman strength due to demonic possession.

However, if we look at bare plurals the facts go in the other direction. Consider (316a, b) with the indefinite subjects replaced by the corresponding bare plural subjects. Then, it does not seem possible to interpret the bare plural subjects existentially.

Diesing (1992) and Kratzer (1995) show that existential readings for indefinite subjects are available only when existential readings for bare plural subjects are also available. The seemingly existential readings in the absence of an actuality entailment in (316) are not true existential readings but are instead instances of specific indefinite readings, which are always available.[79] Therefore, from the absence of existential readings with bare plural subjects in the absence of an actuality entailment, we conclude that existential readings are only available when there is an actuality entailment.

It follows from the above conclusion that instances of momentary/ short-lived ability (cf. 316) and instances of more long term abilities are not truly distinct with respect to genericity/ILP-hood. This is a welcome result since in the languages which mark the perfective/imperfective distinction on ability modals, long term abilities and momentary abilities are both realized by the same form: an ability modal in the imperfective.

4.2. Back to compositional derivation

The goal is to try and connect the two readings of *be able to*: *managed to* and *had the ability to*. We assume that underlying the two readings of *be able to* is one predicate which combines with different operators to yield the two readings. Let us call this predicate *ABLE*[80] and assume that *ABLE* is a non-stative implicative verb with a conventional implicature somewhat similar to *manage to*.

The fact that in the episodic past in English, and past perfective in Bulgarian, Catalan, Greek, Hindi etc., there is an actuality entailment does not come as a surprise, but instead follows from our assumption that *ABLE* is an implicative verb.

(317) John managed to eat the pizza → John ate the pizza.

The predicate embedded under the implicative verb is evaluated with the matrix tense specification.

The interpretive possibilities available to bare plural/indefinite subjects of *be able to* follow from our analysis of *be able to* as a non-stative predicate. The English past tense allows for both episodic and generic readings of non states but the English present only allows for generic readings.

(318) a. Firemen lift heavy cinder blocks. (only gen.)

b. Firemen lifted heavy cinder blocks. (gen./∃)

These are exactly the possibilities found with *be able to.* (cf. 319a, c) (and also with *managed to* cf. 319b, d).

(319) a. Firemen are able to lift heavy cinder blocks. (only gen.)

b. Firemen were able to lift heavy cinder blocks. (gen./∃)

A similar pattern is found with *managed to.*

(320) a. Firemen manage to lift heavy cinder blocks. (only gen.)

b. Firemen managed to lift heavy cinder blocks. (gen./∃)

The conventional implicature discussed in Sect. 4.1 survives to the actual ability attribution. It is somewhat odd to attribute trivial abilities such as the ability to lift one's finger. However, such ability attributions become perfect if the context makes it clear that they are not trivial for the person to whom the ability is being attributed.

(321) a. # Timmy is able to breathe.

b. Timmy had a terrible car accident as a result of which he lost control over most of his muscles. Thankfully, he is able to breathe.

So (321a) is odd because its conventional implicature is not satisfied. Once the conventional implicature is satisfied, as in (321b), the oddness vanishes.

The LFs in (322) show how the ability attribution reading ('had the ability to') of *be able to* is derived. In Bulgarian, Catalan, Greek, Hindi-Urdu etc. the *Gen* operator is contributed by the semantics of imperfective aspect.

(322) a. (In those days,) A fireman was able to eat five apples.
LF: Past (*Gen* (*ABLE* (eat-5-apples)) (**fireman**))

b. A fireman is able to eat five apples.
LF:*Gen* (*ABLE* (eat-5-apples)) (**fireman**)

The absence of actuality entailments is not surprising since the LFs in (322a, b) do not entail the corresponding generic sentences. Also consider the fact that (323a) does not entail (323b).

(323) a. John manages to sit through a Chinese opera.

b. John sits through a Chinese opera.

It may be argued that the analysis offered here suffers from the problem of too strong truth conditions. Even in the generic, implicative verbs need some verifying instances. Consider (323), which can presumably not be said unless John has on some occasions sat through a Chinese opera. Such a need for verifying instances seems absent with *be able to*. However, this problem only arises if we identify the semantics of *be able to* with the semantics of *manage to*. Not all generic sentences require verifying instances. Consider the following examples of generic sentences from Carlson (1995).

(324) a. This machine crushes up oranges and removes the seeds.
b. The Speaker of the House succeeds the vice president.
c. Sally handles the mail from Antarctica.

Carlson (1995) notes that (324a-c) seem quite possibly to be true even under circumstances where the corresponding episodes do not ever take place. So it is not the case that all generic sentences require verifying instances. Lawler (1973) and Dahl (1975) have noted that generic sentences can have two quite different readings as can be seen in (325).

(325) John drinks beer.

(325) has a 'universal'/ habitual reading under which John habitually drinks beer (all the time or on the relevant drinking occasions) and an 'existential'/ dispositional reading that says that John does not object to drinking beer. It seems the dispositional reading does not require verifying instances while the habitual reading does. Why it is the case that (322a, b) only involve the dispositional flavor of genericity, however, still remains to be explained. The fact that ABLE combines with the dispositional flavor of genericity is what makes us think of it as 'modal'.[81]

Another potential problem is Hackl (1998)'s analysis of ability *can/is able to* as an individual level predicate. Hackl argues that with respect to compatibility with locative modifiers and quantificational adverbs, *can/is able to* pattern with individual level predicates like *be intelligent* and not derived generics like *eat pizza*. The point he makes about *can/is able to* can be extended to non actuality entailment *was able to* also. Are his analysis and the current treatment of the *had the ability to* ability attribution as a derived generic compatible? We believe that an assimilation is possible if we adopt an analysis of individual level predicates developed by Chierchia (1995). Chierchia's analysis treats individual level predicates as *derived* inherent generics. He derives the differences that exist between individual level predicates and transparently derived generics from lexical properties of individual level predicates and not from an appeal to the underived status of individual level predicates. The relevant features of his account can be adopted by us to give the derived generics in (322) individual level properties. So there is no contradiction between assuming individual level properties for the 'had-the-ability-to' ability attribution and treating it as a derived generic.

A problem that we do not have much to say about at this point is the incompatibility of *be able to* with the progressive. If *be able to* is a non-stative predicate, we expect it to be compatible with the progressive. This expectation is not met.

(326) * John was being able to eat the pizza.

While we do not have an explanation, we note that other implicative verbs are not perfect in the progressive either.

(327) ? John was managing to eat the pizza.

Further, the ill-formedness of (326) may very well be related to the fact that adjectives in English do not generally occur in the progressive.

5. Conclusions and extensions

We saw that sentences like 'John was able to lift the truck' are ambiguous between a 'managed to' implicative verb-like reading which asserts that John actually lifted the truck and says nothing about John's abilities and a 'had the ability to' reading which asserts that John has the ability to lift the truck. The implicative verb reading is only available in non-generic (perfective) environments, while the 'had the ability to' reading is only available in generic (imperfective) environments. In his 1971 paper on implicative verbs, Karttunen writes:

> If the quarterback in 46a (= 'In the last game, the quarterback was able to complete only two passes') did not in fact complete two passes, it is very improbable that anybody would regard 46a as true. (pg. 355)
> What remains to be explained , however, is why *be able* and other similar verbs in contexts like 46a, *which I am unable to describe in any general way*, must be interpreted as giving not only a necessary but a sufficient condition for the truth of the embedded sentence. (pg. 356) (emphasis: RB)

We have provided a general characterization of the environments where *be able to* behaves like an implicative verb and made a proposal of crosslinguistic import as to why *be able to* behaves the way it does. According to our proposal, the source of the modality in *able to* is a dispositional *Gen* operator. The connection between modality and the *Gen* operator receives independent support from recent proposals advanced in Menéndez-Benito (2005) concerning the licensing of free choice items by the *Gen* operator. Across the languages where the actualization pattern is found, we see that free choice items are only licensed in the absence of the actuality entailment.

(328) Hindi-Urdu: *koi-bhii* 'some-also/even' is a free choice/negative polarity item

a. perfective aspect, actuality entailment: free choice item not licensed

Rohit* **koi-bhii *mez uṭhaa sak-aa*
Rohit.m some-also table lift can-Pfv.MSg

b. habitual aspect, no actuality entailment: free choice item licensed

Rohit **koi-bhii** *mez uṭhaa sak-taa hai*
Rohit.m some-also table lift can-Hab.MSg be.Prs.Sg

'Rohit can/is able to lift any table.'

Within our proposal, the actuality entailment arises in the absence of the *Gen* operator and it is the absence of a suitable licenser that leads to the ungrammaticality of (328).

We conclude this chapter with the observation that the correlation between actualization and perfective aspect is not limited to *be able to*. The pattern found with *be able to* is replicated with a number of other lexical items and even constructions in a number of languages. Some such cases are discussed below.

5.1. *Trop* 'too' and *assez* 'enough' in French

One case that has been recently discussed and analyzed in some detail by Hacquard (2005/2006) involves that of *too* and *enough* in French. Hacquard notes that *too* and *enough* interact with aspect in a manner similar to ability modals. In the perfective, the infinitival complement of *assez* 'enough' is entailed. With *trop* 'too', the negation of the infinitival complement is entailed.

(329) *assez* 'enough' (from Hacquard (2005))

a. perfective: actualization: P-enough to Q $\rightarrow$ Q

Jean a été assez rapide por s'enfuir (#mais il ne s'est
Jean was-Pfv enough quick to escape but he NE SE-is
pas enfui).
Neg escaped

'Jean was quick enough to escape (#but he didn't escape).

b. imperfective: no actualization

Jean était assez rapide por s'enfuir (mais il ne s'est
Jean was-Impfv enough quick to escape but he NE SE-is
pas enfui).
Neg escaped

'Jean was quick enough to escape (but he didn't escape).

(330) *trop* 'too' (from Hacquard (2005))

a. perfective: actualization: too-P to Q $\rightarrow \neg$ Q

Jean a été trop lent por s'enfuir (#mais il s'est enfui).
Jean was-Pfv too slow to escape but he SE-is escaped

'Jean was too slow to escape (#but he still escaped).

b. imperfective: no actualization

Jean était trop lent por s'enfuir (mais il ne s'est
Jean was-Impfv too slow to escape but he SE-is escaped
pas enfui).

'Jean was too slow to escape (but he still escaped).

Hacquard analyzes *trop* 'too' and *assez* 'enough' as implicatives that presuppose a sufficient and necessary condition for the realization of their complements.

(331) a. $[\![\text{enough}]\!] = \lambda x \lambda P_{dest} \lambda Q_{st}.$
$[\iota d : \forall w \in \text{Acc}(w^*).Q(w) \leftrightarrow P(d)(x)(w)]P(d_1)(x)(w)$
where $d_1 = \iota d : [\forall w \in \text{Acc}(w^*).Q(w) \leftrightarrow P(d)(x)(w)]$

b. $[\![\text{to}]\!] = \lambda x \lambda P_{dest} \lambda Q_{st}.$
$[\iota d : \forall w \in \text{Acc}(w^*).\neg Q(w) \leftrightarrow P(d)(x)(w)]P(d_1)(x)(w)$
where $d_1 = \iota d : [\forall w \in \text{Acc}(w^*).\neg Q(w) \leftrightarrow P(d)(x)(w)]$

When *too/enough* clauses combine with the Generic operator, their presuppositions get accommodated into the restriction of the Generic operator. Hacquard (2005) argues that this forces us to only consider situations that depend in the adjective. The actual situation may be different and that is why we don't get an actualization.

5.2. Other cases of actualization

A number of languages have predicates that do not entail their complements ordinarily but do do so in perfective aspect. This is the case with the Greek verb *epitrepo* 'permit'[82] and the French predicate *have the courage to.*

(332) the Greek verb *epitrepo* 'permit' (Iatridou p.c.):

a. perfective → actualization

i karta mu epetrepse na xrisimopiiso tin vivliothiki
this card me allow.pst.pfv use the library
'This card permitted me to use the library.' (I used the library)

b. imperfective → no actualization

i karta mu epetrepe na xrisimopiiso tin vivliothiki
this card me allow.impfv use the library
'This card permitted me to use the library.' (I don't have to have used the library)

(333) French 'have the courage to' (from Hacquard (2005))

a. perfective → actualization

Jean a eu le courage de parler à Marie, #mais il ne lui
Jean had-Pfv the courage to talk to Marie but he NE her
a pas parlé.
has Neg talked
'Jean had the courage to talk to Mary, #but he didn't talk to her.'

b. imperfective → no actualization

Jean avait le courage de parler à Marie, mais il ne lui
Jean had-Impfv the courage to talk to Marie but he NE her
a pas parlé.
has Neg talked
'Jean had the courage to talk to Mary, but he didn't talk to her.'

This is of course the same pattern found with ability modals. It remains to be seen whether the proposals advanced for ability modals will extend to these cases too.

A final case comes from Hindi-Urdu. This case differs from the ones surveyed so far in that it does not involve an overt lexical item on which the meaning alternation can be based. Hindi-Urdu has a passive that can only appear in NPI-licensing environments, referred to variously as the inabilitative/capacitative passive. This passive has a modal meaning which indicates inability. Like the ability modal, the inabilitative passive also displays the by now familiar actualization pattern.[83]

(334) Hindi-Urdu inabilitative passives:

a. perfective → actualization

Mona-se ek ghanṭ-me sirf tiin ghaṛe banaa-ye
Mona-Instr one hour-in only three pots.m make-Pfv.MPl
gaye
Pass-Pfv.MPl
'Mona was only able to make three pots in an hour.' (only three pots were made)

b. imperfective → no actualization

Mona-se ek ghanṭ-me sirf tiin ghaṛe banaa-ye
Mona-Instr one hour-in only three pots.m make-Pfv.MPl
jaa-te hẼ
Pass-Hab.Mpl be.Prs.Pl
'Mona is only able to make three pots in an hour.' (no pots necessarily made)

Assimilating the inabilitative passive to the other instances of the actualization-aspect correlation would involve identifying a covert element with implicative semantics that combines with the generic operator to yield modal semantics. Given that the inabilitative passive can only appear in NPI-licensing environments, we are independently driven to postulate a null polarity item as a part of the makeup of the inabilitative passive and it is plausible that this null element has the appropriate semantics. A proper investigation of this line of enquiry will, however, have to wait for another occasion.

Notes

1. I would like to thank Amy Rose Deal for giving me comments on this chapter.
2. A different, and very interesting, approach to some of the questions raised in this chapter is developed in Pesetsky and Torrego (2001).
3. This idea is suggested in Kjellmer (1975).
4. Jones (1991) proposes that the differences between the two constructions can be reduced to differences in the size of the complement the matrix *be* takes. He claims that the *be* in the propositional *be* construction embeds a purpose clause, which he proposes is a VP. The modal *be* construction embeds a full clause, which he claims is a CP.
5. The idea that the source of the infinitival modality lies in the infinitival complementizer has been around since Bresnan (1972). Developments of this idea include Stowell (1982), Pesetsky (1991), and Portner (1992, 1997) among others.
6. Giannakidou (1998) analyzes *na* as a subjunctive marker. The point being made here is independent of any particular analysis of *na* though.
7. It is possible though that *humming the song* in (33) involves covert structure which is responsible for licensing the PRO.
8. It should be noted though that this problem only arises for Kayne if he assumes, as I do, that *of* is a dummy case-marker. If *of* is serving some other function, the argument that the ungrammaticality of (34b) is due to a failure of case-licensing loses its force.
9. Since this text was originally written, the proper analysis of relative clauses has been the subject of much work. An extended version of the discussion in this section can be found in Bhatt (2002); reactions to Bhatt (2002) include Sharvit (2004), Bhatt and Sharvit (2005), Heycock (2005), , and Hulsey and Sauerland (2006) among others. Sauerland (2003) provides an insightful overview.
10. This analysis is so ubiquitous that it is not clear where it originated. Quine (1960) seems to suggest it. It is assumed in Montague (1970), Partee (1975), Chomsky (1977), Jackendoff (1977) among many others.
11. The conclusion reached in Sauerland (1998/2003) and Bhatt (2002) is that we need both the raising and the matching analysis but that the head-external analysis is unavailable.
12. Thanks to Roumi Pancheva for suggesting this argument to me.
13. The obligatory relative-clause internal interpretation of the head NP that we see in (42) and (43) is not a general property of relative clauses. There exist cases where the head NP must be interpreted in a relative-clause external position (cf. 61).
14. See Grosu & Landman 1998 for an attempt. They provide a mechanism that allows us to extract the 'substance' out of the degree description. Their approach provides the desired result in cases where there is no intervening modal operator. It is not clear to me that it would extend to cases like (46), where we are not talking about any specific books that Gina needs or any particular drinks that Mary can take.
15. The idea that reduced relatives have a PRO subject has been around since at least Burzio (1986). It has also been adopted in recent work by Iatridou, Anagnostopoulou, and Izvorski (1998/2001), among others.
16. The point holds definitionally and rather vacuously for subject infinitival relatives. If the object is relativized, the result is grammatical but we no longer have a subject infinitival

relative. Further, the resulting object is not a reduced relative.

17. Things are actually trickier. At least at LF, we do not seem to need a C^0 to form a predicate. Perhaps we need a C^0 to do pre-LF A'-movement - something that English relative clauses need.
I am also side-stepping the question of what kinds of clausal constituents can work as nominal modifiers. It seems to be the case that having the semantic type $< e,t >$ is a necessary but not sufficient condition for a predicate to function as a nominal modifier.

i. a. the boy [$_{CP}$ who/that ate the apple]
b. the boy [$_{PrtP}$ eating the apple]
c. *the boy [$_{VP}$ eat the apple]
d. the apple [$_{PrtP}$ eaten by Loida]
e. the firemen [$_{AP}$ available]
f. *the firemen [$_{VP}$ be available]
g. the students [$_{AP}$ likely to win the prize]
h. *the students [$_{VP}$ seem to win the prize]
i. the books [$_{IP}$ to be read]
j. the books [$_{PP}$ on the table]

The predicates in (i.c,f,h) seem to be the right semantic type ($< e,t >$) but it is not a sufficient condition to permit the relevant clausal constituents to function as nominal modifiers. Loosely speaking, it seems that to be able to function as a nominal modifier, a clausal constituent has to be 'adjectival/nominal' - a property shared by prepositional and adjectival phrases and participial and infinitival clauses. It is not obvious how this notion of 'adjectival/nominal' can be characterized precisely.

18. The motivations for the PRO A'-movement approach within the Heim & Kratzer (1998) are (i) to bind variables, and (ii) to create a (propositional) scope domain. The system proposed by Heim & Kratzer only requires *movement*, not A'-movement. If there are functional projections within the PrtP to which the PRO can A-move, then the motivations for the PRO A'-movement disappear.

19. The discussion here is not meant to rule out a direct predicate-formation analysis of all infinitival predicates. That seems to be needed for cases like (i) (cf. Clark (1990), Landau (1999)) and may also be available (redundantly) for reduced relatives in general. The local movement of PRO in (i) is meant to indicate that the infinitival complements are interpreted as predicates. Landau's analysis proposes that a class of non-psychological adjectives take predicates as complements. His analysis is independent of the particular method of creating this predicate (Direct Predication or PRO A'-movement).

i. a. The book$_i$ is available [PRO$_1$ [t$_1$ to be read]].
b. The story is not fit [PRO$_1$ [t$_1$ to be printed]],
c. The volcano is ready [PRO$_1$ [t$_1$ to erupt]]. (on the non-psych reading of *ready*)

20. I assume that finite/object infinitival relative clauses always involve a relative operator - overt or covert. A relative operator is necessary to trigger A'-movement. Note that there

are many languages which lack null relative operators altogether. The obligatory presence of the relative pronoun can be used to explain why the ungrammaticality of **the the book John read*. The presence of a covert relative operator (which I presume is a D^0) would block the appearance of *the*. The ungrammaticality of **the the book John read* also plausibly follows from a certain set of assumptions about semantic composition which would require the sister of the predicate *John read* to be of type $< e,t >$. *the book* would be of type e or $<< e,t >,t >$ leading to ungrammaticality due to a type-mismatch.

21. Lasnik (1998) argues against A-reconstruction. If we assume with him that there is no A-reconstruction, we would use whatever mechanism he would use to derive the 'reconstructed' readings in cases of A-'reconstruction' with PrtP-internal *A*-chains. The point relevant here is that nothing special needs to be said for reduced relatives.
22. This assumption seems to have been around since at least the time when the DP hypothesis was proposed but it is not clear to me where it originates.
23. There are other ways to explain the ungrammaticality of **the the philosopher reading the Meinong text*: (i) the wrong semantic types explanation - *the* needs some thing type $< e,t >$ as an argument, *the philosopher* is of type e (or $<< e,t >,t >$) and the *the philosopher reading the Meinong text* is of type t. We will get ungrammaticality due to an unresolvable type-mismatch; (ii) the subcategorization of *the* argument: *the* subcategorizes for an NP complement. So * *the the philosopher*.
24. For many speakers, there is a contrast between (69a) and (69b). While a non-modal readings seems to be available to such speakers in both (69a) and (69b), these speakers find (69a) degraded.
25. Kjellmer (1975) also provides examples without a cardinality adjective (cf. i).

    i. a. He is one of the businessmen to hold a top post in America.

       b. Mr. and Mrs. Warburton were among the guests to arrive later.

    He notes that if the explicit partitive is removed, the non-modal reading of (i) disappears.

    ii. a. They are the businessmen to hold a top post in America.

        b. Mr. and Mrs. Warburton were the guests to arrive later.

    I agree with the contrast between (i) and (ii) but I find the examples in (i) less than perfect. The acceptability of the cases in (i) suggests that partitivity plays an important role in licensing the non-modal interpretation, perhaps a role that supersedes the effect of cardinality adjectives.
26. Wells (1947) suggested that 'first'/'tallest' composed with the infinitival clause as a discontinuous constituent first and then composed with the noun (cf. Berman 1973 for developments of this idea, and Smith 1964 for a related idea). The analysis proposed here captures the insight behind Wells' suggestion.
27. The judgements regarding (80) do not seem to be very strong. Some speakers are able to get the photography reading where it is indicated not to be available.
28. Sharvit and Stateva (2002) present arguments against a movement approach to superlatives. The *in-situ* approach outlined in sect. 8 would be compatible with their proposal.
29. As with superlatives, ordinals/*only* can be provided either a one-argument or a two-argument semantics. LFs are provided for both options in this section. An explicit two-argument semantics for Ordinals will be provided in sect. 6.1. Sect. 7 discusses some

proposals for the semantics of *only* and also provides a translation of the two-argument semantics for Ordinals into a one-argument semantics.

30. In the LFs indicated, for simplicity I am ignoring the fact that adverbial *only* can only associate with material in its surface syntactic scope cf. the absence of association of *only* with focus in '**John**$_F$ only introduced Bill to Sue'.
31. Stump (1981), and Larson (1998) respond to the conceptual problem raised by the movement postulated in (97) by arguing that the reading in question can be derived without moving *occasional* out of the subject NP. They propose that *occasional* and the determiner together form a quantifier that quantifies over events. This allows the NP-internal *occasional* to in effect behave as an adverbial modifier without actually moving out of the NP.
32. Kayne (1998) challenges the assumption that *only* in (98) forms a constituent with the NP *one banana*. He derives the configuration that we are claiming obtains at LF in overt syntax.
33. This is modulo the fact that *many/few/two* permit non-modal readings when an overt partitive *one of* is present. See sect. 1. At present, I have nothing to say about why the presence of an overt partitive makes the difference it does.
34. If we did not have the rule of Predicate Modification, numerals, *few* and *many* would have to be provided the higher semantic type $<< e,t >,< e,t >>$. We would then need another way of stating the difference between adjectival modifiers like numerals, *few* and *many* and focus-sensitive ordinals, superlatives, and *only*.
35. It is also possible that (105b, c) point to a deeper problem with the statement of the licensing configuration. In Appendix B (sect. 8), a version of the licensing configuration will be discussed which does not involve a relative clause-internal interpretation of the head NP. This alternative proposal is able to provide a very straightforward account for the deviance of (105b, c).
36. The readings shown here in the paraphrases are the 'association with focus'/ comparative readings. In addition, there are other readings e.g. the 'absolute' reading where the telescope is the most expensive out of the telescope Joan owns/Mary owns/ the telescopes in the world.
37. The superlative NP has been put in post-copular position deliberately because otherwise definite NPs seem to be sometimes capable of setting time reference independently of matrix tense. This is shown by the acceptability of 'John's first restaurant is a laundromat now.'/'The fugitives are all in jail now.' cf. Enç (1986), Musan (1997).
38. Here, the assumption is implicitly made that unbound time variables are freely available in the semantic computation. In sect. 6.5.1, we will question this assumption and adopt a more restrictive theory of access to unbound time variables.
39. This comports well with the result arrived at in Bhatt (2002) that the grammar does not make available the head-external analysis. Only the head-raising and the matching analysis are options. Both have a relative clause internal copy of the external head.
40. A way out would be to assume that the past tense in the relative clause is not there at LF i.e. the past tense is just a surface past, somewhat like the past tense in Sequence of Tense contexts. However SOT past-tense deletion/ presence of the dummy past tense seems to be (i) optional, and (ii) dependent upon the presence of a sufficiently local Past Tense. However, if SOT past-tense deletion is what happens in (136), it cannot be optional. If it were, the simultaneity reading would have been available. Further, the simultaneity

effects do not seem to be contingent upon the presence of a matrix past tense. Therefore, we do not pursue the SOT past-tense deletion line further.

41. It is assumed in (140a) that a relative clause, which is a predicate of individuals (semantic type $< e,t >$), can become a predicate of times and individuals ($< T, < e,t >>$). I switch to a one-argument treatment of *first* in (140b). I believe that a two-argument treatment is possible given certain assumptions about the interpretation of reconstructed material in relative clause.
42. Strictly speaking, in the cases we have considered so far, it is the head NP and the event associated with the relative clause that have to be simultaneous.
43. The difference between formalizing past tense as referential and formalizing it as indexical does not seem to have any empirical consequences for the current discussion.
44. In this we seem to go against the 'intuition' that in *the first man to solve the problem, first* is somehow more closely related to the infinitival clause. According to this 'intuition', we start off with *the [first [to walk on the moon]] man*, which is then followed by extraposition of the infinitival clause to the edge of the NP. I think that 'intuition' has something to it - the relationship between *first* and the infinitival clause is a very local one (cf. the licensing configuration). However, it does not have to exclude the head NP and we show here that it cannot exclude the head NP.
45. Following Schlenker (1998), I will ignore the definiteness in the postcopular NP. Schlenker points out that postcopular definite predicative nominals do not pattern with non-predicative definites with respect to triggering of presuppositions (cf. i).

    i. a. My cat is not grey. (presupposition: I have a cat)
       b. This is not my cat. (no presupposition regarding whether I own a cat)
       c. The chancellor of Lusitania is not blonde. (presupposition: Lusitania has a unique chancellor)
       d. Olafur is not the chancellor of Lusitania. (no presupposition regarding whether Lusitania has a chancellor)

    See also Doron (1984) for related observations.
46. Note that the 'interpret the head NP relative clause externally also' strategy works for finite relative clauses like *the first 80 yr. old who walked* . Here too we find the **Head NP** and the **Simultaneity** requirement. The analysis requires obligatory interpretation of the head NP in a position below the RC-internal past tense (thus paralleling obligatory interpretation of the head NP under the covert perfect in *the first 80 yr. old to walk on the moon*.).
47. There seems to be some variation with respect to speaker judgements with respect to the overt perfect (cf. i).

    i. Bill is the first 80 yr. old to have walked on the moon since the war.
       Speaker Group 1: Bill is 80 yrs. old, He has walked on the moon since the war.
       Speaker Group 2: Bill is 80 yrs. old. He was 80 yrs. old when he walked on the moon.

    The difference between the two groups of speakers seems to be that the latter group have a simultaneity requirement with an overt *have* also. Group 1 allows for simultaneity, it just does not require it. Both groups impose the head requirement. At present, I have no explanation for the judgements of the people in Group 2.
48. A definition of *first* that works with intervals is provided in (i).

i $first(P) = \lambda x \lambda T[P(x,T) \wedge \forall y[[x \neq y \wedge \exists T'.P(y,T')] \Longrightarrow F(P)(x) < F(P)(y)]]$
where
$F(P)(x) = \iota \lambda t.[\exists T'[t \in T' \wedge P(x,T')] \wedge \forall t' < t.\neg \exists T''[t \in T'' \wedge P(x,T'')]]$ (F extracts the moment at which P begins to hold for x)

49. The first exception is that of questions like *When to leave?*, *Who to talk to at the party?*, *How to solve this problem?*, *Where to get gas?*, and *What to do now?*. These cases are judged as marginal and have a limited distribution. The other exception is provided by questions like *Why worry?*. These cases are analyzed as grammaticized modal rhetorical questions in Bhatt (1998b). We do not delve into the distribution of matrix *wh*-infinitivals in this work. Reis (2003) provides a discussion and analysis of comparable data involving *wh*-infinitivals in German.
50. Judgements indicated are Lahiri's. I find these examples significantly worse - ???/*.
51. It also follows if we assume that c-selection is very local. As a result when predicates c(ategorially)-select for questions, they do not look at the features of the complement of C^0.

    i a. ... Pred $[_{CP}\ [_{C}$+wh$]\ [_{TP}$ ... infinitival clause ...]]

    b. ... Pred $[_{CP}\ [_{C}$+wh$]\ [_{TP}$ ... finite clause ...]]

    All that subcategorization for a question involves is selection of a `[+wh]` CP. The finiteness of the TP complement of C^0 is not available to be selected.
    This locality story has the following potential problem: we need to assume that the finiteness or non-finiteness of a complement is not represented in C^0. This assumption does not hold if C^0s come marked for the features of their complements e.g. *for* could have the `[+inf]` feature, or if there is T-to-C movement and the information of the embedded IP is available to the matrix predicate.
52. It should be kept in mind that finite indicative clauses, and questions, finite or infinitival, are CPs. Non-question infinitival clauses may or may not be CPs. The cases where it proves to not be feasible to provide a unified analysis of $PR_{-wh,+inf}$ and $PR_{+wh,+inf}$ may be cases where the complements of $PR_{-wh,+inf}$ and $PR_{+wh,+inf}$ are categorially distinct. In such cases, the non-interrogative infinitival complement selected by $PR_{-wh,+inf}$ would be a maximal projection smaller than a CP while the complement of $PR_{+wh,+inf}$ has to be a CP.
53. *learn* also has an ability reading. Cf. *Magnus is learning to play the drums* and the almost synonymous *Magnus is learning how to play the drums*. With infinitival question complements, the ability reading only surfaces with *how*. I will abstract away from this reading here.
54. In the examples that we will be considering, epistemic readings are easily available when the propositional complement of the modal is interpreted in the past. Deontic readings are easily available when the propositional complement of the modal is interpreted in the future.
55. The definitions of $\Box_D$ and $\Diamond_D$ are simplified versions of the definitions in Kratzer (1991). The simplification resides in the assumption that there is always a set of 'closest' worlds that the Ordering Source can pick out. Kratzer's definition does not make this assumption and is consequently more complicated.
56. This method of picking out the most informative answer assumes that there will always be a unique smallest subset of the question denotation that entails everything else in the

question denotation. It is not clear to me that the unique existence of such a set can be taken as given. In case there is no unique subset, we will need to develop a more refined notion of 'most informative'.

57. Cases of the following sort are also relevant. Suppose talking to Magnus, Herb, and Penna (= G_1) satisfies the goals. Talking to Daniel, Stefan-Árni, and Baldur (= G_2) also satisfies the goals. However, talking to G_1 plus someone from G_2 and vice versa does not lead to popularity. The intuitions seems be less than sharp at this point. It seems clear that a more detailed answer would qualify e.g. one that provided the information that talking to Magnus, Herb, and Penna and not talking to anyone else satisfies the goals. This would be predicted by the proposed system. But it is not so clear that just knowing the talking to Magnus, Herb, and Penna part is insufficient. As it stands, the proposed system has problems with the issue of how fine-grained the specification of the action needed to satisfy the goals needs to be. We will address this issue in Section 3.2.
58. Hintikka (1976) observed the absence of exhaustivity in infinitival questions. He took the existence of non-exhaustive readings with infinitival readings to show that non-exhaustive readings are available in general. As much subsequent literature from Karttunen (1977, fn. 4) to Lahiri (1998) has noted, such an assumption predicts the existence of readings that do not, in fact, exist.
59. I would like to relate this absence of the exhaustivity requirement to the absence of indicative tense. This would fit with attempts to assimilate non-indicative complements of *have/be* to questions (cf. Izvorski 1998). One notable hurdle to the assimilation of these non-indicative complements to questions is the absence of exhaustivity effects in such complements. If we connect exhaustivity to the presence of indicative morphology and not to the semantics of questions, the project of assimilating non-indicative complements of *have/be* to questions becomes more feasible.
60. (240) has another reading where *to eat well* means *to eat heartily*. The intuition is that in that case there is no *could* paraphrase. The intended reading for (240) is one where *to eat well* means *to eat good food*.
61. The term *factive* is being used somewhat loosely here. The factive *know* that we are familiar with takes propositional complements while the *know* we are talking about here takes question complements. Factivity is a notion that is used to classify predicates that take propositional complements. A predicate $\mathscr{P}$ that takes a propositional complement p is factive iff $\mathscr{P}(p)$ presupposes the truth of p i.e. both $\mathscr{P}(p)$ and $\neg\mathscr{P}(p)$ entail p. This definition has to be extended to cover predicates that take question complements. The following extension captures the extended sense in which the notion of *factive* is used:

    i. A predicate $\mathscr{P}$ is factive with respect to its question complement $\mathscr{Q}$ iff
    $\mathscr{P}(\mathscr{Q}) \iff \mathscr{P}(TrueAnswer(\mathscr{Q}))$

    By this extended definition, both *know* and *tell* are factive when they take question complements.
62. In the general case, one has to exclude the 'superset'/less-informative answers. Uniqueness is required to hold only after all the non-maximally informative answers have been eliminated.
63. This relaxation of exhaustivity is distinct from the pick-one/mention-some readings discussed in the literature. A pick-one/mention-some answer is a member of the the question denotation. The non-exhaustive interpretation that we are using (cf. 268b) is not a member of the question denotation. A pick-one/ mention-some style relaxation of exhaustivity

seems to be empirically undesirable in *should* questions (cf. i).

i Didda knows who she should talk to at the party.
Context: Didda should talk to Magnus, Herb, and Penna.
true members of $[\![who\ she\ should\ talk\ to\ at\ the\ party]\!] =$
$\{\Box_D(talk(didda,m)), \Box_D(talk(didda,h)), \Box_D(talk(didda,p)),$
$\Box_D(talk(didda,m+h)), \Box_D(talk(didda,h+p)), \Box_D(talk(didda,p+m)),$
$\Box_D(talk(didda,m+h+p))\}$

The intuition is quite clear that if Didda only knows that she should talk to Magnus (or any proper subset of Magnus, Herb, and Penna), (i) is false. If a pick-one/mention-some style relaxation of exhaustivity were available here, we would expect (i) to be judged as true.

64. Overt *should* does not seem to permit deontic ordering sources of the sort permitted with the infinitival modality. So in (i), we do not get the goal of getting gas as an ordering source.

i Didda knows where she should get gas.

If the goal of getting gas could be an ordering source, on a parallel with the non-exhaustive reading in (268), (i) would have a reading on which it would mean something like 'Didda knows where she could get gas'. On this reading it would be equivalent to 'Didda knows where to get gas' on the *could* reading. Such a reading seems to be absent. The non-parallelism between *should* and the covert infinitival modality, though of interest, does not constitute any kind of analytical puzzle. The infinitival modality is not a covert version of *should*.

65. We are assuming that the negation distributes over the plural object i.e. $\neg getGas(h,a+b+c)$ entails $\neg getGas(h,a)$, $\neg getGas(h,b)$, and $\neg getGas(h,c)$. The kind distributivity that we need to assume is also found in examples like (i).

i. I didn't see the children.

(i) entails 'I didn't see any of the children' and not just the weaker proposition 'I didn't see all the children'. If such a covert distributivity is not assumed, the question denotation will not contain any true answers. It is not clear why there is a covert distributivity requirement in negative infinitival questions.

66. Corpus search reveals an unusual infinitival question question with *how*:

i a. "They 're out there on the cutting edge of labor law, on how better to represent your clients," according to Arthur Fox, a TDU lawyer.

b. Short of that, Darrell's elders "should not be too arrogant and think they know how best to run a country," Dr. Mahathir added.

c. The computer, calculating the cards remaining in the deck, sent back messages advising him, based on mathematical probability, how best to play his hand.

The fronted element in these cases does not seem to form a constituent and to a first approximation, the questions seem to mean the same as their counterparts with only the *how* fronted. Like the comparable cases with just the *how* fronted, the cases in (i) do not seem to involve any degree quantification by the *how*.

67. The assumption that traces of degree extraction are interpreted as variables over degrees is made by Cresswell (1976), von Stechow (1984), among many others.
68. The proposal that traces of degree extraction be interpreted as variables over sets of degrees is made by Schwarzschild (1998) reporting joint work with Karina Wilkinson.
69. Hackl & Nissenbaum (1999) note the existence of a systematic class of exceptions to the Determiner Modal Generalization. This class of exceptions involves instruments and is shown in (i).

    i. a. Every pen to write with is in the top drawer.
       b. Most guns to shoot quail with have wooden handles.

    They are exceptional in that the infinitival modality is interpreted as *could* despite the presence of strong quantifiers. This exceptionality does not follow from the proposal advanced here.
70. The feature of the semantics of covert modality that allow for the jump from 'There is a unique book s.t. reading that book leads to goal satisfaction' to 'There is a unique x s.t. reading that book x leads to goal satisfaction' distinguishes this covert modality from overt modality and is in part responsible for why this covert modality, and not overt modality, display variable force effects (cf. i, ii).

    i. a. The/a book (for you) to read (the-should, a-should/could)
       b. The/a book you should read (no effect of determiner on force of modality)
       c. The/a book you could read (no effect of determiner on force of modality)

    ii. a. Hafdis knows [which book to read] (*should*)
        b. Hafdis knows [which book she could read] (no effect of *which*-phrase on modality)
        c. Hafdis knows [which book she should read] (no effect of *which*-phrase on modality)

    The choice of determiner/*wh*-phrase determines whether the infinitival modality can be paraphrased as *could*/*should*. This choice does not have an influence on the interpretation of overt modals however.
71. Martin Hackl (p.c.) has noted that if the derivation of a *should* reading from $\Diamond_{D,\rightarrow}$ based on uniqueness effects relies on the semantics of questions, then it is not clear that it will carry over to relative clauses. This concern is well-taken but it seems to us that the derivations do not make crucial reference to the semantics of questions. The particular derivations did make reference to the Groenendijk & Stokhof style partition induced by the question but we do not see why relative clauses cannot generate the same kind of partition.
72. Giannakidou (2003b) points out the following potential problem with the covert partitivity approach. If strong readings always involve partitivity as suggested here and if partitivity is able to at least marginally license non-modal readings of subject infinitival relative clauses as noted in Chapter 3, then one would expect strong quantifiers to license non-modal readings, counter to fact. At this point, we cannot properly address this problem because we do not understand the role played by partitivity in licensing non-modal readings in the first place.

73. I am assuming that a plural head like *women* contains in its extension both singular and plural women. So if there are only three women in this world a, b, and c, the extension of *women* is the set $\{ a,b,c,a+b,a+c,b+c,a+b+c \}$.
74. In order to properly address the difference between $\Diamond_{D,\rightarrow}$ and $\Diamond_{D}$, we need to decide how the contextual restriction C in the definition of $\Diamond_{D,\rightarrow}$ is set when $\Diamond_{D,\rightarrow}$ appears in an infinitival relative. C made reference to the Groenendijk & Stokhof partition in the case of infinitival questions and we need to determine for relative clauses what the appropriate source of partitions is.
75. Judgements by Sabine Iatridou.
76. In addition to genericity, the Greek imperfective has an additional 'event-in-progress' reading. The Hindi imperfective lacks this reading, which is realized by a separate progressive marker. Because of the absence of the 'event-in-progress' reading, *-taa* the marker of imperfective aspect is sometimes glossed as 'Hab(itual)'.
77. The specialized progressive marker in Hindi-Urdu is in fact incompatible with the ability modal

    i. *#/*Yusuf havaii-jahaaz uṛaa sak rahaa hai*
    Yusuf air-ship fly CAN Prog.MSg be.Prs.Sg
    '*Yusuf is being able to fly the plane.'

78. For a few speakers, (316a, b) are quite odd. For most other speakers, the temporary ability without any accompanying actualization reading is available. We do not have an explanation for this variation in judgements currently.
79. It is instructive to look at the case of ability modals in the present tense. There is no actuality entailment and existential readings for indefinite and bare plural subjects are not available (cf. i).

    i a. A fireman is able to do fifty pushups in a minute.

    b. Firemen are able to do fifty pushups in a minute.

    However, even here, in (ia), the indefinite can be interpreted specifically as in 'a fireman, namely Michael'.
80. At this point, we have already embarked on the project of treating the modal reading as consisting of a non-modal component and a Generic operator. Piñon (2003) criticizes this project and proposes an alternative where the modal reading is underived and the actuality entailment is derived from the modal reading. We do not offer a full comparison of our proposal with Piñon's proposal but it would be fair to say that his proposal does not derive the correlation with aspect that our proposal is based on. Hacquard (2006) presents a sophisticated proposal that derives the correlation with aspect as well as why the correlation holds only for a subclass of modals, namely the root modals. Her proposal departs from mine in that the underlying component is modal to begin with. This also give her a way to handle the problem that dispositional genericity by itself does not yield ability.
81. Angelika Kratzer has pointed out to me that deriving ability from the combination of a non-modal ABLE and dispositional genericity is problematic because it does not seem to be the case that we get ability readings when dispositional genericity combines with other non-modal predicates. See also footnote 80.

82. With the Greek *epitrepo*, the facts are more complicated. If the subject is agentive, there does not seem to be an actuality entailment. Thus the Greek counterpart of *The librarian permitted me to use the library* in the perfective does not entail that I used the library. Similar facts seem to hold for the Italian counterpart of *epitrepo* 'permit'.
83. A closely related variant of the inabilitative passive is the inabilitative unaccusative. Unaccusative predicates in Hindi-Urdu do not allow for the realization of agents but inabilitative unaccusatives, which like their passive counterparts are polarity sensitive, require the realization of an agent as an instrumental *-se* phrase. And like the inabilitative passive, we find the correlation between aspect and actualization.

    i. Hindi-Urdu inabilitative unaccusatives:

        a. perfective → actualization

            Mona-se ek ghanṭ-me sirf tiin ghaṛe ban-e
            Mona-Instr one hour-in only three pots.m make$_{unacc}$-Pfv.MPl
            'Mona was only able to make three pots in an hour.' (only three pots were made)

        b. imperfective → no actualization

            Mona-se ek ghanṭ-me sirf tiin ghaṛe ban-te hẼ
            Mona-Instr one hour-in only three pots.m make$_{unacc}$-Hab.Mpl be.Prs.Pl
            'Mona is only able to make three pots in an hour.' (no pots necessarily made)

References

Anagnostopoulou, Elena, Sabine Iatridou, and Roumi Izvorski
1998 On the morphosyntax of the Perfect and how it relates to its meaning. In: P. N. Tamanji and K. Kusumoto, (eds.), *Proceedings of NELS 28*, Amherst, Massachusetts, GLSA, 1–17.

Anderson, Stephen R.
1972 How to get even. *Language* 48, 893–906.

Baker, Carl L.
1995 *English Syntax*. MIT Press, Cambridge, MA. 2nd Edition.

Beck, Sigrid
1996 Wh-constructions and transparent Logical Form. Ph.D Dissertation, Universität Tübingen, Tübingen, Germany. Distributed SfS Report 04-96.

Berman, Arlene
1973 Adjectives and Adjective Complement Constructions in English. Ph.D Dissertation, Harvard University, Cambridge, Massachusetts.

Berman, Stephen
1991 On the semantics and logical form of Wh-Clauses. Ph.D Dissertation, University of Massachusetts-Amherst, Amherst, Massachusetts. Distributed by GLSA.

Bhatt, Rajesh
1997 Obligation and Possession. In: H. Harley, (ed.), *The Proceedings of the MIT roundtable on argument structure*, MITWPL, Cambridge, MA.

Bhatt, Rajesh
1998 *Argument-Adjunct asymmetries and Rhetorical Questions*. Manuscript of talk presented at NELS29 at the University of Delaware.

Bhatt, Rajesh
2002 The Raising Analysis of Relative Clauses: Evidence from Adjectival Modification. *Natural Language Semantics* 10:1, 43–90.

Bhatt, Rajesh and R. Pancheva
2004 Late Merge of Degree Clauses. *Linguistic Inquiry* 34:3.

Bhatt, Rajesh and Y. Sharvit
2005 A note on intensional superlatives. In: E. Georgala and J. Howell, (eds.), *Proceedings of SALT XV*, 62–79. Cornell University, Ithaca, NY, CLC Publications.

Bolinger, Dwight
1967 Adjectives in English: Attribution and Predication. *Lingua* 18, 1–34.

Brame, Mike K.
1968 *A new analysis of the relative clause: evidence for an interpretive theory*. Unpublished manuscript, MIT.
Brame, Mike K.
1976 *Conjectures and Refutations in Syntax and Semantics*. New York: North-Holland Publishing Co.
Bresnan, Joan
1972 Theory of Complementation in English Syntax. Ph.D Dissertation, Massachusetts Institute of Technology, Cambridge, Massachusetts.
Burzio, Luigi
1986 *Italian syntax: a government-binding approach*. (Studies in natural language and linguistic theory). Dordrecht: Kluwer.
Butt, Miriam
1995 *The structure of complex predicates in Urdu*. Dissertations in Linguistics, CSLI Publications, Stanford, California. Doctoral thesis at Stanford University.
Carlson, Greg N.
1977 Amount Relatives. *Language* 53:3, 520–542.
Carlson, Greg N.
1995 Truth Conditions of Generic Sentences: Two Contrasting Views. In: G. N. Carlson and F. J. Pelletier, (eds.), *The Generic Book*, 176–223. Chicago: University of Chicago Press, Chicago.
Chierchia, Gennaro
1995 Individual-level Predicates as Inherent Generics. In: G. N. Carlson and F. J. Pelletier, (eds.), *The Generic Book*, 176–223. University of Chicago Press, Chicago.
Chomsky, Noam
1965 *Aspects of the Theory of Syntax*, MIT Press, Cambridge, MA.
Chomsky, Noam
1977 On wh-movement. In: P. Culicover, T. Wasow, and A. Akmajian, (eds.), *Formal syntax*, 71–132. New York: Academic Press.
Chomsky, Noam
1993 A minimalist program for linguistic theory. In: K. Hale and S. Keyser, (eds.), *The view from Building 20: essays in linguistics in honor of Sylvain Bromberger*. 1–52. (Studies in Linguistics 24), Cambridge, MA, MIT Press.
Chomsky, Noam
1995 *The minimalist program*. Studies in Linguistics 28, MIT Press, Cambridge, Massachusetts.
Chomsky, Noam
1998 *Minimalist Inquiries: The Framework*. MIT Occasional Papers in Linguistics 15, MITWPL, Cambridge, MA.

Chomsky, Noam and Howard Lasnik
1977 Filters and Control. *Linguistic Inquiry* 8:4, 425–504.

Clark, R. L.
1990 *Thematic Theory in Syntax and Interpretation* (Croom Helm linguistics Series). Routledge, London.

Dahl, Östen.
1975 On Generics. In: E. L. Keenan, (ed.), *Formal Semantics of Natural Language: Papers from a Colloquium Sponsored by the King's College Research Centre*, 99–111.Cambridge: Cambridge University Press.

Dayal, Veneeta
1996 *Locality in Wh-quantification: Questions and Relative Clauses in Hindi.* (Studies in Linguistics and Philosophy 62) Kluwer, Dordrecht.

Diesing, Molly
1992 *Indefinites.* (Linguistic Inquiry Monographs 20) MIT Press, Cambridge, Massachusetts.

Dikken, Marcel den, Richard K. Larson, and Peter Ludlow
1997 Intensional 'Transitive' Verbs and Concealed Complement Clauses. In: P. Ludlow, (ed.), *Readings in the Philosophy of Language.* MIT Press, Cambridge MA.

Doron, Edit
1984 Verbless Predicates in Hebrew. Ph.D Dissertation, University of Texas at Austin, Austin, Texas.

Dowty, David R.
1979 *Word meaning and Montague Grammar: The semantics of verbs and times in Generative Semantics and Montague's PTQ.* (Studies in Linguistics and Philosophy 7). Kluwer, Dordrecht.

Dubinsky, Stanley
1997 Infinitival Relative Clauses in English: An antisymmetric approach to discontinuous constituency. In: A. D. Green and V. Motapanyane, (eds.), *Proceedings of ESCOL 96*, , 82–93. Cornell University, Ithaca, NY, Cornell Linguistics Club.

Elliott, Dale
1971 The Grammar of Emotive and Exclamative Sentences in English. Ph.D Dissertation, Ohio State University, Columbus, Ohio.

Embick, David
1997 Voice and the interfaces of syntax. Ph.D Dissertation, University of Pennsylvania, Philadelphia, PA.

Enç, Mürvet
1986 Towards a Referential Analysis of Temporal Expressions. *Linguistics and Philosophy* 9:4, 405–426.

Farkas, Donka F.
1988 *Intensional Descriptions and the Romance Subjunctive Mood*, Outstanding dissertations in linguistics, Garland, New York.

Fox, Danny
1999 Reconstruction, Binding Theory, and the Interpretation of Chains. *Linguistic Inquiry* 30:2, 157–196.

Fox, Danny
2001 *The Syntax and Semantics of Traces*. Handout of Nov. 2001 talk at the University of Connecticut.

Fox, Danny
2002 Antecedent Contained Deletion and The Copy Theory of Movement. *Linguistic Inquiry* 34, 63-96.

Geisler, Christer
1995 *Relative Infinitives in English*. Acta Universitatis Upsaliensis, Almqvist and Wiksell International, Stockholm, Sweden. Doctoral thesis at Uppsala University 1995 (Studia Anglistica Upsaliensia 91).

Giannakidou, Anastasia
1998 *Polarity Sensitivity as (Non) Veridical Dependency* Amsterdam: John Benjamins.

Giannakidou, Anastasia
2003 Review of 'Covert Modality in non-finite contexts'. *Glot International* 7:1-2.

Grimshaw, Jane
1979 Complement Selection and the Lexicon. *Linguistic Inquiry* 10:2, 279–326.

Grimshaw, Jane
1977 English Wh-Constructions and the Theory of Grammar. Ph.D Dissertation, University of Massachusetts-Amherst, Amherst, Massachusetts. Distributed by GLSA.

Groenendijk, Jeroen and Martin Stokhof
1982 Semantic Analysis of Wh-Complements. *Linguistics and Philosophy* 5:2, 175–233.

Groenendijk, Jeroen and Martin Stokhof
1984 Studies on the Semantics of Questions and the Pragmatics of Answers. Ph.D Dissertation, University of Amsterdam, Amsterdam, The Netherlands.

Grosu, Alex and Fred Landman
1998 Strange Relatives of the Third Kind. *Natural Language Semantics* 6:2, 125–170.

Hackl, Martin
1998 *On the semantics of ability attributions*. Unpublished manuscript, Massachusetts Institute of Technology.

Hackl, Martin and Jonathan Nissenbaum

1999 Variable Modal Force in For-infinitival Relative Clauses. In: A. Giorgi, J. Higginbotham, and F. Pianesi, (eds.), *The Proceedings of The Bergamo Conference on the Syntax and Semantics of Tense and Mood Selection*, Oxford: Oxford University Press (to appear).

Hacquard, Valentine

2005 Aspects of Too and Enough Constructions. In: E. Georgala and J. Howell, (eds.), *Proceedings of SALT XV*, Cornell University, Ithaca, NY, CLC Publications.

Hacquard, Valentine

2006 Aspects of Modality. Ph.D Dissertation, Massachusetts Institute of Technology, Cambridge, Massachusetts.

Hamblin, C. L.

1973 Questions in Montague English. *Foundations of Language* 10, 41–53.

Heim, Irene

1987 Where does the Definiteness Restriction Apply? Evidence from the Definiteness of Variables. In E. Reuland and A. ter Meulen, (eds.), *The Linguistic Representation of (In)definiteness*, 21-42. Cambridge, MA.: MIT Press.

Heim, Irene

1992 *Kroch, Cinque, Negative Islands*. Handout from 1992 lecture, Massachusetts Institute of Technology.

Heim, Irene

1995 *Superlatives: A case study on the division of labor of syntax and semantics*. Unpublished manuscript, Massachusetts Institute of Technology.

Heim, Irene and Angelika Kratzer

1998 *Semantics in Generative Grammar* (Blackwell Textbooks in Linguistics 13). Oxford: Blackwell.

Heycock, Caroline

2005 On the interpretation of adjectival modifiers and relative clauses. *Natural Language Semantics* 13:4, 359–382.

Heycock, Caroline and Antony Kroch

1999 Pseudocleft Connectivity: Implications for the LF Interface Level. *Linguistic Inquiry* 30:3, 365–397.

Higginbotham, James

1993 Interrogatives. In: K. Hale and S. J. Keyser, (eds.), *The view from Building 20: essays in linguistics in honor of Sylvain Bromberger*, 195–22. (Studies in Linguistics 24), Cambridge, MA, MIT Press.

Hintikka, Jaakko

1976 *The semantics of questions and the questions of semantics: case studies in the interrelations of logic, semantics, and syntax* (Acta philosophica Fennica 28.4). Amsterdam: North-Holland Publishing Co.

Horn, Larry
1969 A Presuppositional Analysis of only and even. In: A. Davison, G. M. Green, and J. L. Morgan, (eds.), *Papers from the Fifth Regional Meeting of the Chicago Linguistics Society*, 98–107. Chicago: Chicago Linguistics Society.

Huddleston, Rodney
1971 A problem in Relative Clause Reduction. *Linguistic Inquiry* 2:1, 115–116.

Hudson, Richard A.
1973 Tense and Time Reference in Reduced Relative Clauses. *Linguistic Inquiry* 4:2, 251–256.

Hulsey, Sarah and Uli Sauerland
2006 Sorting out Relative Clauses. *Natural Language Semantics* 14:2, 111–137.

Iatridou, Sabine, Elena Anagnostopoulou, and Roumi Izvorski
2001 Some Observations about the Form and Meaning of the Perfect. In: M. Kenstowicz, (ed.), Ken Hale*: A Life in Language*, 189– 238. Cambridge, MA,

Izvorski, Roumi
1998 Non-indicative Wh-Complements of Possessive and Existential Predicates. In: P. N. Tamanji and K. Kusumoto, (eds.), Proceedings of NELS 28, Amherst, Massachusetts, GLSA, 159–173.

Jackendoff, Ray
1977 *$\bar{X}$-Syntax: A study of Phrase Structure*, MIT Press, Cambridge, MA.

Jones, Charles Foster
1985a Syntax and Thematics of Infinitival Adjuncts. Ph.D Dissertation, University of Massachusetts-Amherst, Amherst, Massachusetts. Distributed by GLSA.

Jones, Charles Foster
1985b Two BE + Infinitival Constructions. In: J. Goldberg, S. Mackaye, and M. Wescoat, (eds.), *Proceedings of the West Coast Conference on Formal Linguistics* vol. 4, 112–122. Palo Alto, Ca: Stanford University.

Jones, Charles Foster
1991 *Purpose Clauses - syntax, thematics, and semantics of English purpose constructions* (Studies in Linguistics and Philosophy 47), Kluwer, Dordrecht.

Karttunen, Lauri
1971 Implicative Verbs. *Language* 47:2, 340–358.

Karttunen, Lauri
1977 Syntax and Semantics of Questions. *Linguistics and Philosophy* 1:1, 3–44.

Karttunen, Lauri and Stanley Peters
1979 Conventional Implicature. In: C.-K. Oh and D. A. Dinneen, (eds.), *Presupposition, Syntax and Semantics* 11, 1–55. New York: Academic Press, Harcourt Brace Jovanovich.

Kayne, Richard S.
1993 Towards a modular theory of auxiliary selection. *Studia Linguistica* 47, 3–31.

Kayne, Richard S.
1994 *The Antisymmetry of Syntax* (Linguistic Inquiry Monographs 25) Cambridge, MA.: MIT Press.

Kayne, Richard S.
1998 Overt vs. Covert Movement. *Syntax* 1:2, 128–191.

Kjellmer, Göran
1975 Are relative infinitives modal? *Studia Neophilologica* 47, 323–332.

Klein, Wolfgang
1994 *Time in language*. London: Routledge.

Ko, Heejeong
2005 Syntax of Why-in-situ: Merge into [Spec,CP] in the overt syntax. *Natural Language and Linguistic Theory* 23:4, 867–916.

Kratzer, Angelika
1981 The notional category of Modality. In: H.-J. Eikmeyer and H. Rieser, (eds.), *Words, worlds, and contexts: new approaches in word semantics* (Research in text theory 6), 38–74. Berlin/New York: de Gruyter.

Kratzer, Angelika
1991 Modality. In: Arnim von Stechow and Dieter Wunderlich, (eds.), *Semantics, An International Handbook of Contemporary Research*. Berlin/New York: de Gruyter.

Kratzer, Angelika
1995 Stage-level and Individual-level Predicates. In: G. N. Carlson and F. J. Pelletier, (eds.), *The Generic Book*, 125–175. Chicago: University of Chicago Press.

Kuroda, Shigeyuki
1968 English Relativization and Certain Related Problems. *Language* 44, 244–266.

Ladusaw, William A.
1980 Polarity Sensitivity as inherent scope relations. Ph.D Dissertation, University of Texas at Austin, Austin, Texas.

Lahiri, Utpal
1991 Embedded Interrogatives and Predicates that embed them. Ph.D Dissertation, Massachusetts Institute of Technology, Cambridge, Massachusetts. Distributed by MIT Working Papers in Linguistics.

Lahiri, Utpal
1998 *Questions and Answers in Embedded Contexts*. Book manuscript, University of California-Irvine.

Landau, Idan
1999 Elements of Control. Ph.D Dissertation, Massachusetts Institute of Technology, Cambridge, Massachusetts.

Larson, R.
1985 Bare-NP Adverbs. *Linguistic Inquiry* 16:4, 595–621.

Larson, Richard K.
1998 Events and Modification in Nominals. In: *Proceedings of SALT VIII*, 145-168. Cornell University, Ithaca, NY, Cornell Linguistics Club.

Larson, R, and Sunguen Cho
1998 *Temporal Modification and the DP Hypothesis*. Paper presented at the UCLA Tense Workshop, May 16-20, 1998.

Lasnik, Howard
1998 Some Reconstruction Riddles. In: A. Dimitriadis, H. Lee, C. Moisset, and A. Williams, (eds.), *Proceedings of the 22nd Annual Penn Linguistics Colloquium* (Penn Working Papers in Linguistics 5.1). University of Pennsylvania Linguistics Club, Philadelphia, PA.

Lawler, John M.
1973 *Studies in English generics*. (University of Michigan Papers in Linguistics 1). Ann Arbor, Michigan: University of Michigan Press.

Lees, Robert B.
1960 *The Grammar of English Nominalizations*. The Hague: Mouton.

Lees, Robert B.
1961 The Constituent Structure of Noun Phrases. *American Speech* 36, 159–168.

Linebarger, Marcia
1980 The Grammar of Negative Polarity. Ph.D Dissertation, Massachusetts Institute of Technology, Cambridge, Massachusetts. Distributed by MIT Working Papers in Linguistics.

McCoard, Robert W.
1978 *The English Perfect: Tense-Choice and Pragmatic Inference* (North-Holland Linguistic Series 38). Amsterdam: North-Holland Publishing Co.

Menéndez-Benito, Paula
2005 The Grammar of Choice. Ph.D Dissertation, University of Massachusetts-Amherst, Amherst, Massachusetts.

Michaelis, Laura A. and Knut Lambrecht
1996a The Exclamative Sentence Type in English. In: A. E. Goldberg, (ed.), *Conceptual Structure, Discourse, and Language*, 375–390. CSLI Publications, Stanford, California.

Michaelis, Laura A. and Knud Lambrecht
1996b Toward a construction-based theory of language function: the case of Nominal Extraposition. *Language* 72:2, 215–247.

Montague, Richard
1974 English as a formal language. In: R. H. Thomason, (ed.), *Formal Philosophy: Selected papers of Richard Montague*, Yale University Press, New Haven, CT. Originally published in 1970.

Musan, Renate
1997 *On the temporal interpretation of noun phrases*. Outstanding dissertations in linguistics, Garland, New York.

Partee, Barbara
1975 Montague Grammar and Transformational Grammar. *Linguistic Inquiry* 6.

Pesetsky, David
1991 *Zero Syntax: Vol.2 - Infinitives*. Unpublished manuscript, MIT, available from web-site.

Pesetsky, David
1997 Some Optimality Principles of Sentence Pronunciation. In: P. Barbosa, D. Fox, P. Hagstrom, M. J. McGinnis, and D. Pesetsky, (eds.), *Is the Best Good Enough?*, MITWPL, Cambridge, MA.: MIT Press.

Pesetsky, David and Esther Torrego
2001 Some Consequences of T-to-C Movement. In: M. Kenstowicz, (ed.), *Ken Hale: A Life in Language*, 355–426. Cambridge, MA: MIT Press.

Piñon, Chris
2003 Being Able To. In: G. Garding and M. Tsujimura, (eds.), *The Proceedings of the 22nd West Coast Conference on Formal Linguistics*, 384–397. Somerville, MA: Cascadilla Press.

Portner, Paul
1992 Situation Theory and the Semantics of Propositional Expressions. Ph.D Dissertation, University of Massachusetts-Amherst, Amherst, Massachusetts. Distributed by GLSA.

Portner, Paul
1997 The Semantics of Mood, Complementation, and Conversational Force. *Natural Language Semantics* 5, 167–212.

Pullum, Geoffrey, and Deirdre Wilson
1977 Autonomous syntax and the analysis of auxiliaries. *Language* 53:4.

Quine, Willard Van Orman
1960 *Word and object, Studies in communication*. Cambridge, MA.: MIT Press

Reis, Marga
2003 On the Form and Interpretation of German Wh-Infinitives. *Journal of Comparative Germanic Linguistics* 7:2, 155–201.

Rizzi, Luigi
1990 *Relativized Minimality* (Linguistic Inquiry Monographs 16). MIT Press, Cambridge, MA.

Romero, Maribel
1997 The Correlation between Scope Reconstruction and Connectivity Effects. In E. Curtis, J. Lyle, and G. Webster, (eds.), *Proceedings of the 16th West Coast Conference on Formal Linguistics*, 351-366. Palo Alto, CA: CSLI.

Rooth, Mats E.
1985 Association with Focus. Ph.D Dissertation, University of Massachusetts-Amherst, Amherst, Massachusetts. Distributed by GLSA.

Rooth, Mats E.
1992 A Theory of Focus Interpretation. *Natural Language Semantics* 1:1, 75–116.

Ross, J. R.
1964 A partial grammar of English superlatives. Unpublished manuscript of M.A. Thesis, University of Pennsylvania.

Ross, John R.
1984 Inner Islands. In: C. Brugmann and M. Macaulay, (eds.), *Proceedings of the 10th Annual Meeting of the Berkeley Linguistics Society*, 258-265. Berkeley, California: Berkeley Linguistics Society.

Rullmann, Hotze
1995 Maximality in the semantics of Wh-Constructions. Ph.D Dissertation, University of Massachusetts-Amherst, Amherst, Massachusetts. Distributed by GLSA.

Rullmann, Hotze and Sigrid Beck
1997 Reconstruction and the interpretation of which-phrases. In: *Proceedings of the Tübingen Workshop on Reconstruction*, 215–249. Universität Tübingen.

Safir, Ken
1999 Vehicle Change and Reconstruction in A'-chains. *Linguistic Inquiry* 30:4, 587–620.

Sauerland, Uli
1998 The Meaning of Chains. Ph.D Dissertation, Massachusetts Institute of Technology, Cambridge, Massachusetts. Distributed by MIT Working Papers in Linguistics.

Sauerland, Uli
2003 Unpronounced heads in relative clauses. In: K. Schwabe and S.Winkler, (eds.), 205–226. *The Interfaces: Deriving and Interpreting Omitted Structures*. Amsterdam: John Benjamins.
Schachter, Paul
1973 Focus and Relativization. *Language* 49:1, 19–46.
Schlenker, Paul
1998 *Pseudoclefts and Concealed Questions*. Unpublished manuscript of generals paper, MIT.
Schwarzschild, Roger
1998 *Interval Analysis of Clausal Comparatives*. handout of talk given at MIT, reports work done with Karina Wilkinson.
Seuren, Pieter A. M.
1978 The Structure and Selection of Positive and Negative Gradable Adjectives. In: D. Farkas, W. M. Jacobsen, and K. W. Todrys, (eds.), *Papers from the Parasession on the Lexicon*, 336–346.. Chicago: Chicago Linguistics Society.
Sharvit, Yael
1999 Functional Relative Clauses. *Linguistics and Philosophy* 22:5, 447–478.
Sharvit, Yael
2004 Two Reconstruction Puzzles. University of Connecticut ms. To appear in the *Proceedings of the Brown Conference on Direct Compositionality*.
Sharvit, Yael and Penka Stateva
2002 Superlative Expressions, Context, and Focus. *Linguistics and Philosophy* 25, 453–505.
Smith, Carlota
1964 Determiners and Relative Clauses in a Generative Grammar of English. *Language* 40:1, 37–52.
Stowell, Tim
1982 The Tense of Infinitives. *Linguistic Inquiry* 13, 561–570.
Stump, Gregory T.
1981 The Interpretation of Frequency Adjectives. *Linguistics and Philosophy* 4:2, 221–257.
Szabolcsi, Anna
1986 Comparative Superlatives. In: N. Fukui, T. R. Rapoport, and E. Sagey, (eds.), *Papers in Theoretical Linguistics* (MIT Working Papers in Linguistics 8), 245–265. Cambridge, MA: MIT.
Thalberg, Irving
1969 Austin on Abilities. In: K. T. Fann, (ed.), *Symposium on J. L. Austin*, 182–205. New York: Humanities Press Inc.

Vergnaud, Jean-Roger
1974 French Relative Clauses. Ph.D Dissertation, Massachusetts Institute of Technology, Cambridge, Massachusetts.
Wells, Rulon S.
1947 Immediate Constituents. *Language* 23:2, 81–119.
Williams, Edwin S.
1975 Small Clauses in English. In: J. P. Kimball, (ed.), *Syntax and Semantics* (Syntax and Semantics 4), 249–273. New York:. Academic Press, Harcourt Brace Jovanovich.

Index